AMERICA'S ENERGY

AMERICA'S ENERGY
A Small Book About
a Big Problem

Robert Harston
RGH Associates, LLC
Denver, Colorado

RGH Associates, LLC
Denver, Colorado

Cover design: Bryan Harston
Chart design: Derrick Sample

Photo credits:
> Front Cover: David Gilder, ©2004 iStockphoto.com/dgilder
> Back Cover: Mustafa Deliormanli, ©2009 iStockphoto.com/deliormanli
> Author Photo: Karin Gallo

Harston, Robert
America's Energy: A small book about a big problem / Robert Harston

> p. cm.

> Includes bibliographical references and index

ISBN-13: 978-1-4505987-5-0
Library of Congress Control Number: 2009913040

First printing: December 2009

Printed and bound in the United States of America

DEDICATION

This book is dedicated to the men and women of America's energy industries, for one simple reason: you make everything *work*. Everything we have, everything we use, and everything we need ... they all depend on energy. From coal mines and oilfields; into power plants and refineries; through the electric power grid and gas pipelines; to the wall sockets in our living rooms and the gas stoves in our kitchens ... it's the energy industries that make it happen.

This work is further dedicated to the educators who taught, and continue to teach, those men and women the skills, arts, and sciences they need to find, produce, and deliver that energy to America. The training of young minds for the future of the energy industry continues at numerous fine institutions across the country and the world. Two of the very finest are the two I know best: The University of Oklahoma and Colorado School of Mines.

But above all, I dedicate this book to my wife, Sharyl. Your encouragement and confidence in me helped bring this book to life.

Bob Harston
November 2009

CONTENTS

PREFACE

Seven hundred billion dollars. Every year.

That's real spent money. Together, we — you, I, and the rest of America — paid that much for imported crude oil ... *just last year*. We paid nearly that amount the year before. We will almost certainly pay that much (if not more) next year.

We, as a nation, are exporting our cash. That money goes largely to nations and peoples who are not necessarily our best friends in the world. And a significant portion of that money — to the tune of billions — goes to avowed enemies of America. This outflow of cash buys the U.S. our annual imports of crude to meet America's energy needs. In my view, neither of these flows — not the export of so much cash, nor the import of so much oil — is a good thing.

Hopefully this book will provide some unvarnished facts on the energy problem; its causes, and some potential solutions. It may also refute some of the "fantasy" solutions which — while certainly exciting — simply won't happen anytime soon. In this book you'll find chapters devoted to energy sources, supply and demand, uses, logistics, and politics. We'll also discuss the relative practicality and promise of emerging technologies ... both real and surreal.

This book is not a technical treatise. It won't solve our energy problems, but it may — by opening up for discussion this vital topic — help inform the discussion, enhancing our ability to think realistically about where we are now.

It's clear, because the author knows more than many about the energy business and unbiased because I don't have any skin-in-the-game ... except our nation's energy future. I don't represent the energy industry, nor do I represent any other industry, government, import or export enterprises, politicians, or political parties.[1]

I also present some options that might help us resolve our energy shortfalls without the political spin or rhetoric of the various industries involved in the usual discussions. The only advocacy in this book is reality with regard to energy: where we are, and where we'll likely be in twenty years.

I have no particular axe to grind with this book, nor do I aspire to be a best-selling author. Do I have biases? Certainly ... as does every person who has lived long enough to form his or her own opinions. But in the following pages I will endeavor to point out viewpoints that are opinion-based; otherwise, we're dealing with facts.

On Climate Change: A Genteel Rant

This book is, as titled, about America's energy. In today's political environment, however, an overarching energy issue is, and will continue to be, the realities and causes of climate change. We'll get more into that in our "Environment" portion of the book, but it's worth a few opening remarks here:

Yes, climates are changing globally. I do not now, nor have I ever, disagreed with those who assert the Earth's climate is changing. Were it *not* changing, it would probably be for the first time in several billions of years ... as the Earth's climate is a in a relatively constant state of change. (Remember: dinosaurs once grazed in

---

[1] In the interest of full disclosure: I spent about 30 years in the oil & gas information services business, first in business development and ultimately in senior management. I own a few energy company stocks, and have some mutual fund holdings which probably own energy-related stocks in their portfolios. (None of them check with me on investment strategy.) I also served on the Board of small fabricating firm involved in wind tower manufacturing.

tropical jungles … beneath what are now wheat fields in Kansas and ski areas in the Rocky Mountains.)

For as long as climatic records have been kept, they have noted some fluctuation in climates. Some years it's minor, some years it's (relatively) large. There are numerous recognized drivers of Earth's constant historical climate changes: variations ("wobble") in the earth's elliptical orbit around the Sun; subtle shifts the tilt of the earth on its axis; variations in the sun's radiation and cosmic ray emissions (what we laymen may call "sunspots").

Where I take issue with the mainstream hype is with the argument that mankind's activities on earth have (or even *could*) materially impact the earth's climate. More specifically, that carbon dioxide (CO_2) emissions are the cause of global warming. Although I do comment further on my own skepticism, this book is certainly not a scientific reference, and does not offer a solution to the argument.

Nevertheless, whether carbon or other factors are to blame, current legislation, EPA regulations, emotions, and publicity guarantee that new carbon regulation will have a direct impact on every watt of electricity and every gallon of gasoline we use.

Whether it happens through "Cap and Trade" legislation, greater restrictions on the uses of oil, gas, and coal, mandated "alternative" energy sources, or other draconian regulatory actions; our lives and the cost of living will be changed. For believers and non-believers alike, every business investment or land development decision will be weighed with regard to emissions of greenhouse gases (GHGs), one of which is CO_2.

Where my skepticism lies, in part, is that 95% of all GHGs are naturally occurring water vapor, mostly evaporation from the oceans. Of the remaining 5%, about 3.5% is CO_2. The rest are nitrous oxide, methane and other gases. Of the CO_2 component, ⅔ is naturally occurring, man influencing the other ⅓ of that 3.5%. Many scientists tell me that man's activity is the cause of global warming; others say it is not. Yet that conclusion is so counterintuitive that it seems illogical.

While I'm not a scientist, I am a college graduate with some significant experience (in industry, and on the planet generally). Like most people, when presented with conflicting scientific evidence, I tend to filter the conflicting information based on my own experiences, values, and beliefs. When something seems so illogical as to be fantastic, I tend to believe the version that best fits my own logic. From my reading, most people tend to do the same.

This doesn't mean I'm right or wrong about global warming, but it does mean that I — only one of many skeptics — have yet to be convinced. (As we move along, I would remind us all that in the 1970's, the concern *du jour* was "global cooling" ... under which we were all going to freeze to death.)

Interestingly, just as we go to press, all manner of red flags are appearing on the horizon, casting doubt over the factual-basis-versus-bias of important climatology data upon which many world governments – including our own – rely in formulating policy.

Certainly, the jury is still out (if, in fact, it has even convened yet). But it appears, nonetheless, that data which did not conform to expected/planned research findings may have been altered or concealed.

The issue surrounds the East Anglia University's Climate Research Unit (CRU), a globally-recognized provider of climate data and a contributor to the UN's International Panel on Climate Change (IPCC) whose series of three reports, starting in 1990, triggered much of today's frenzy regarding carbon in the atmosphere.

The IPCC purported to have some 2,500 scientific experts concurring on these publications. But according to author Ian Plimer, there were actually fewer than 1,700 "authors" to those reports, and of that number, some 1,200 were, in fact, political and/or environmental activists, not scientists.[2]

---

[2] (Plimer 2009) See footnote 27, below, and Bibliography pages.

WHERE OUR ENERGY COMES FROM; WHERE IT GOES

Like most of the world, the U.S. relies primarily on five distinct energy sources for our needs. Crude oil, coal, natural gas, nuclear and hydroelectric (water-generated) are the leading fuel/power sources used to generate electricity and manufacture fuels for transportation and industrial uses. Indeed the world's energy infrastructure after over a century of development, investment, and use is almost totally oriented to three carbon-based fuels: coal, oil, and natural gas. Even in those parts of the world that use nuclear power for electricity, the principal transportation fuels are still derived from crude oil or natural gas.

As we delve into the supply and demand issues of worldwide energy consumption it is easy to get confused by terms, products and measurements which may confound those not familiar with the various energy industries' nomenclature. For example, sometimes the data is/are presented in BTUs, sometimes in kilowatt hours, sometimes in tons and sometimes dollars. I'll try to normalize and/or define these for clarity and comparison as we go. We'll keep this information general and approximate without getting into the minutiae (which are important, but aren't really the point of this project).

This analysis will address supplies, production, and usage of: oil, coal, natural gas, nuclear, hydroelectric and some alternatives, for the generation of electricity, transportation, and heating. There will not be many precise numbers in this book ... rather they will be ballpark figures. These numbers change (literally) day-to-day. The price of crude oil may move up, down, or both up *and* down, up to five dollars or more either way, in any given 24-hour period. The price of coal is also somewhat unstable at the present time, although not as volatile as crude oil has been of late.

Let's look at the different fuel/power sources and their relative contribution to the overall energy picture.

Crude Oil

For decades, the United States has been a "net importer" of crude oil. In other words, we use more than we produce — *lots* more — so we have to import crude oil to make up the difference.

How much? Well, let's begin with a little simple math:

Oil we **consume**		20 million barrels/day
Oil we **produce**	−	5 million barrels/day
Oil imports we **need**	=	15 million barrels/day

Again, that's fifteen *million* barrels imported. *Every day*. Forty years ago — in the 1970s — America produced about 6 million barrels/day. At the time, that was roughly half our average daily consumption of ±13 million barrels.

PRICES. This book includes many calculations based on the per-barrel price of crude. For ease of calculation, we'll assume oil costs \$100/barrel. Right now it's about two-thirds of that, but within the past eighteen months it's approached \$150. The world's price of crude oil had roughly tripled in the preceding 18 months, but then fell by more than 60%. From its low of about \$30, it doubled at least once (within the span of few months) and, as of this writing, is once again back down. By the time you read this, it will have changed, one way or another. (We'll talk about some "whys" about oil prices later.)

PRODUCTS. Lots of products come from a barrel of crude oil. Most crude goes directly to a refinery, where it is separated into any number of products. These include: liquefied petroleum (LP) gases, such as propane and butane; gasoline (most of which fuels passenger cars); diesel fuel (used principally for trucks, buses, trains, and large industrial engines); kerosene (most of which becomes jet fuel); home heating oil; lubricants, such as motor oil and grease; asphalt and tars, used for road surfacing, roofing, etc.; naphtha and other solvents; and feed stocks (raw ingredients) used to make plastics, chemicals, fertilizer, etc.

So, how much of any given product can we get from one barrel of oil? That depends upon several factors: the quality of the oil itself, the type of refinery facility that receives it, and the demands of the marketplace on any given day (*e.g.*, demand for heating oil is highest in the fall and winter; demand for gasoline goes up during summer vacation season).

For this discussion, let's focus on something most of us purchase regularly: gasoline. The industry-standard unit for measuring oil production, refinery throughput, transport, sale, and consumption is the "barrel" (abbreviated as "bbl.") A barrel of oil represents a volume of 42 U.S. gallons.[3, 4]

---

[3] Although it is certainly the world standard used in accounting for oil produced, sold, transported, and stored, the modern 42-gallon "oil barrel" doesn't exist. Spend as much time as you like rummaging around oil fields, refineries, or gas stations. You won't come across a 42-gallon container ... not these days, anyhow. The "barrel of oil" is really just an arbitrary unit of measure.

[4] The history of this seemingly-odd standard dates back to the early oil days in Pennsylvania. In the mid- to late-19th Century, many sizes of wooden cask — most holding between 40-45 gallons — were used to sell and transport oil. This was problematic for everyone. Sellers would use whatever size barrels they could find, so any given shipment might contain several barrel sizes, and it could be difficult to account for every gallon. Buyers, on the other hand, could not readily verify the volume of a closed barrel, nor view its contents, in the field. This led to confusion in the industry as to how oil should be measured and priced.

So in August of 1866, an industry group known as the Council of Producers decided they would measure and price oil by the *gallon*, rather than by the barrel:

> We, therefore, mutually agree and bind ourselves that from this date we will sell no crude by the barrel or package, but by the gallon only. An allowance of two gallons will be made on the gauge of each and every 40 gallons in favor of the buyer.

(*Derrick's Hand Book of Petroleum.* Volume 1, p. 77. Oil City, PA: Derrick Publishing Company, 1898.) While this meant producers would sell oil by the *gallon*, they still needed barrels — actual wooden casks, mind you — to package and ship it. This was before the days of pipelines and ocean-going oil tankers. Producers needed containers that could be

(-more-)

But these 42 gallons of crude don't translate directly into 42 gallons of gasoline. Only about 60% (25 gallons) of one refined barrel will be refined into gasoline, with the remainder yielding a mix of the other petroleum products mentioned above. (This 60% number is typical, but varies widely.)

QUALITY. Because crude oil is an organic substance formed naturally underground, the oil produced from one well may be quite different from the oil produced from a different well (or field, or region, or continent). There are many qualities of crude oil, some easier to refine than others. At the simplest level, oil is graded according to its relative weight and viscosity ("light" versus "heavy") and its relative sulfur content ("sweet" crude is low in sulfur; "sour" crude is high in sulfur). The thinner/lighter crudes are most desirable and valuable because they yield more of the "light-end" (higher-priced) products — gasoline, kerosene, LP gases) for sale to consumers. Likewise, sweet crude is more valuable than sour crude, because the sour product requires more refining to yield a saleable product.

Extremely thick (heavy) oils, which are in plentiful supply in some places around the world (especially Canada's western provinces), can be as thick as tar. Such oil is extremely difficult and expensive to produce, transport, and refine, because the oil won't flow out of a drilled well (as will thinner oils). Sandstone formations containing those tar-like deposits are often (and appropriately) referred to as "tar sands." With the high price of crude in recent years, the production of oil from tar sands has moved from the laboratory to the field, and has indeed approached economic viability. But as the price of crude has dropped, these developmental projects — which typically employ surface mining-like recovery

---

handled by dock workers and transported in any conveyance, from a river barge to a one-horse cart. Wooden barrels, ranging in size from 40 to 45 gallons, were ideal ... and readily available. (Derrick Publishing 1898)

Buyers took for granted the two-gallon "allowance" on every 40 gallons sold; eventually this led to the "bbl. of 42 gallons," as adopted by the Council for the standard *pricing* unit of oil in 1872 ... although the oil itself was still measured in gallons.

techniques, such as open pit mining — are being slowed, minimized, or shut down. (In the "coal" section following, we look at below-ground coal gasification technologies; some speculate that these techniques might also prove applicable to Canada's "tar sand" deposits in the future.)

When you think about it, it's actually surprising how readily available, reliable, and easy-to-use crude oil-derived products are in our society. Why? Because crude oil is really pretty nasty stuff. It's sticky, gooey, dirty, flammable, hard (and expensive) to find, and it requires lots of energy and noxious chemicals to process it into something useful. But the worldwide energy system is designed around it, so you and I don't have to deal with the messy issues. At least not directly.

INFRASTRUCTURE. Crude oil is a ubiquitous and highly useful source of energy in the world's energy supply. Its products power cars and trucks all over the world. The global energy infrastructure is hugely oriented to carbon fuels. Oil, once found and produced, enters an interconnected system of product distribution and handling which links the wellhead to the gas tank. This system might very well be the world's most extensive, capital intensive, far-reaching, and efficient product infrastructure. It's truly amazing that billions of barrels of oil move safely and reliably through this system of pipelines, tanker ships, refineries, and end-product distribution systems worldwide ... reaching billions of motorists via local fuel pumps.

And this present-day, existing, functional infrastructure is one of my key points on this topic:

The worldwide petroleum distribution complex, representing over a century of development and trillions of dollars invested, cannot — logically or practically — be replaced overnight ... much less abandoned.

Crude is the key to the U.S.' enormous outlay of cash to import oil and meet our energy requirements. In 2008, we imported some 13 million barrels per day (which, at $150 per barrel, gets you to my opening figure of $700 billion per year). If the price of crude goes up $5/barrel, our import bill goes up $75 million *per day*. Right now, crude is around $60 a barrel, but I wouldn't bet on it being that low for too long (see "OPEC" below). The consensus seems to be that crude will stay in the $50 range for a year or so then begin to move up at $10 or so per barrel per year for the next year or two. My personal guess is $75 a barrel in 2010. Even at these more modest prices, our oil imports bill still represents hundreds of billions of annual wealth transfer.

One final note: Crude oil is, for better or worse, the primary driver of the historical rise of the Middle East influence in global political and financial arenas. The 2007 estimates of recoverable world crude oil reserves totaled just over 1,300,000,000,000 barrels (that's one-point-three *trillion*); more than all the oil produced in the last century. Of those, some 70% belong to member nations of the cartel known as the Organization of the Petroleum Exporting Countries, or "OPEC." While OPEC includes non-Middle Eastern nations, the majority of OPEC's membership (and aggregate reserves) are from that part of the globe. (More about OPEC and its impact further on.)

Coal

Coal is the principal fuel for electric power generating plants in the U.S. Roughly half of all our electrical power comes from coal-fired power plants. Coal is recovered by traditional mining techniques: underground tunneling, and by strip- or surface-mining techniques. There are significant environmental impacts from those processes, both from surface mining and from burning the coal.

The U.S. produced some 1.2 billion tons of coal in 2006; about 90% of which was burned to generate electricity. We have huge coal reserves in the U.S., some 263.8 billion tons recoverable via existing mining methods, and as those technologies get better over time the "recoverable" figure will likely rise. As with oil, there are varying qualities of coal, both in terms of its energy content, and its relative cleanliness when burned (the quantity and type of emissions produced). Our highest-quality coal has traditionally come from the eastern part of the U.S., although production is declining in that region. Wyoming is currently the leading coal-producing state, with a relatively-low-sulfur coal being economically mined with surface techniques. There are other significant coal deposits in the western states.

CLEAN COAL. Like all organic fuels (and, indeed, all life on this planet), coal is carbon-based. Burning coal releases carbon (primarily as CO_2) and other impurities which, when released into the atmosphere, raises environmental issues. (More on this topic further along.) But the U.S. has massive reserves of coal, and we rely heavily upon it for electric power. We simply cannot summarily stop using coal to generate power without extraordinary impacts on our nation's economy and our standard of living.

Recognizing the importance of balancing our energy needs and environmental concerns, the coal industry continues to develop — and now has in place — some "clean coal" technologies. The goal of these technologies is to reduce emissions from coal-fired power plants by capturing more and more of the noxious emissions produced when coal is burned. Clean coal processes continue to evolve, and promise greater reductions in carbon emissions in the future.

One clean coal method is to convert the coal into a high-purity, smokeless fuel using a centuries-old technique known as "coking." Using massive airtight ovens, coal is "distilled" by heating it to a very high temperature. In the absence of oxygen, the coal itself approaches ignition temperature but does not burn. Rather, the heat drives off volatile impurities (in gaseous form) from the solid coal mass. These impurities — which include water vapor, methane and other useful fuel gases, plus some noxious gases — are not released into the atmosphere, but instead are captured, cooled, separated, and treated.

This process yields clean-burning gas fuel for power generation, and prevents the emission problems that occur when the coal mass itself is burned. The remaining mass of unburned coal is known as "coke," which has very high energy content, and — since most smoke-generating constituents have been burned away — is prized as an essentially-smokeless fuel. Coke has other industrial uses too (*e.g.*, steelmaking). Future technologies may enable us to effectively and cleanly utilize that huge potential resource for power generation as well.

There are other similar and interesting methods of converting coal into different forms. Coal can be "gasified," converted into a natural gas-like fuel suitable for many applications. It can also be liquefied into fuels suitable for use in car and truck engines.[5] Recent innovations in coal gasification are being tested and commercially implemented. These include the burning of coal *in situ* (below ground, where it formed), without mining.

In this process, coal seam energy is harvested by drilling multiple well bores into the coal seam. At one end of the seam, one borehole is used to ignite the coal, while air is pumped into neighboring holes to support combustion and keep the seam burning. As the coal burns, it gives off heat and gases. The gases are forced away from the heat as combustion progresses through the coal seam. As these heated gases expand, pressure in the formation increases, and the gases move away from the seam via the path of least resistance ... which, in this case, is one or more additional boreholes drilled at the *other end* of the targeted area of the coal seam. The heat and gases are captured, and can be used at the wellhead to generate power.

Many of the noxious gases from coal burning are left behind underground. Any remaining emissions can be filtered as the gas is extracted, thus mitigating the carbon emissions usually associated with burning coal above ground. This could signal a major change in coal technology and potentially enhance its attractiveness as a more environmentally acceptable major fuel source.

---

[5] These processes were largely developed in Europe at the first of the 20th Century, but were first exploited on a national scale — with modest success — by Nazi Germany during WWII. During the war, the Axis had major shortages of fuel for their military vehicles, but had plenty of coal.

While commercially-effective sequestration of carbon from coal is purportedly still 10 years away, test projects are now under way.

There are still objections — environmental, for the most part — to the use of coal as the major source of power generating plant fuel. But coal, is nevertheless, by far the single largest energy source for electric power generation in the nation.

Natural Gas

Natural gas is our cleanest-burning fossil fuel, and therefore a superior fuel from an environmental standpoint. Gas is also a highly-efficient heating source, much more so than heating oil and other heavy hydrocarbons.

Natural gas is measured in cubic feet ("CF"). That's how it's measured at the wellhead and through pipelines, and how it is (typically) sold to you and your neighbors by your local utility supplier.[6] In larger quantities, gas volumes may be measured in thousands, millions, billions, or even *trillions* of cubic feet. (In this context, the letter "M" represents one thousand, just as it does in roman numerals. Thus, "5 MCF" means five *thousand* cubic feet, not "one million" as one might infer.) Millions are thus abbreviated as "MM" (thousand-thousand), so 5 million cubic feet is written as "5 MMCF." *Billions* of cubic feet are shown as the more intuitive "BCF"; one thousand BCF makes one trillion cubic feet ("TCF").

Natural gas production and consumption in the U.S. is around 20 BCF per day. Of that, roughly 30% is used in gas-fired power plants to generate electricity. Some 20% of the natural gas used in the U.S. goes directly into private homes, where millions of Americans burn gas in their central heat furnaces, water heaters, gas cook

---

[6] *Terminology Alert*: As used throughout this book, the word "gas" refers to *natural* gas, the methane-rich hydrocarbon produced from underground wells. This should not be confused with the common usage of "gas" as a nickname for gasoline. (To further confuse things: one form of natural gas, called Compressed Natural Gas, or "CNG," serves as an efficient and clean-burning fuel for car, truck, and bus fleets where private refueling networks are available.)

---

tops/ovens, clothes dryers, fireplaces, and/or outdoor grills. Another 30% of the natural gas used in the U.S. goes to other industrial and commercial uses.

The U.S. has huge reserves of natural gas – quite a bit more today, in fact, than were known *and producible* just a few years ago. In 2006, total "proved" reserves[7] in the U.S. amounted to some 212 TCF. At our present rates of consumption (20 BCF/day, or 7.3 TCF/year), those reserves would supply the U.S. for approximately 30 years.

But in the past few years, new drilling and recovery technologies have finally made it cost-effective to produce vast *potential* gas reserves, which have long been known, but – until now – too costly to extract.[8] This has added to our known reserve figures, in that they provide a means for us to extract gas from materials (and locations) we could not have before. Recent major discoveries have also increased our known reserves, including shale-bed gas discoveries in Louisiana and the Appalachian geological basin, and in the Bakken shale found across the northern plains.

How much more gas? In its December 2008 biennial report, the Potential Gas Committee (PGC), considered the authority on natural gas reserves in the U.S., reported that domestic reserves have risen 58 percent in just the last four years. According to the report, the U.S. possesses a total resource base of 2,074 TCF.[9] This is the highest resource evaluation in the Committee's 44-year history. In a July 18 2009 story about the PGC's report, *The New York Times* said "The finding raises the possibility that natural gas could

---

[7] Gas reserves are "proved" when there is at least a 90% chance of recovery of the gas in a known gas-producing formation underground. "Known" and "Potential" reserves exist where the location and quantity of in-place gas has been established to a substantial certainty, but the ability of present-day technologies to extract said gas is not fully known.

[8] Principally, gas found in underground coal and shale formations.

[9] (Potential Gas Committee 2008) At present-day consumption rates of 7.3 TCF/year, that gas resource base could potentially supply the U.S.'s natural gas needs for nearly 300 years.

emerge as a critical transition fuel that could help to battle global warming."[10]

Natural gas prices are fairly moderate at present, well below the price of oil on an energy-equivalent basis.[11] Gas is a desirable fuel that burns cleanly and has far less environmental impact than coal, and significantly less than crude oil-derived fuels as well. From an energy-equivalent standpoint, 5,000 cubic feet of gas (5 MCF) equals about one barrel of oil. So natural gas "should be" about $15/MCF if oil was $75. The main reason it is typically cheaper, compared to the price of oil, is because we have a great deal of it right here at home, and it's less expensive to process and transport (see below).

One significant cost advantage of natural gas is that it does not have to be "refined" in the same way crude oil does, in order to produce a useful end product. Natural gas requires processing, but it's relatively little compared to that required for crude oil. We import significant quantities of natural gas from Canada and Mexico, but this is mostly due to limited production and distribution capacity from our existing wells, not a shortage of domestic reserves. As we drill more producing wells, and construct more distribution pipelines, the volume of imports should drop.

We also import some natural gas from large overseas producers, which is delivered in liquid form. Liquefied Natural Gas (LNG) is piped from producing wells and fields to terminals, where it is chilled to some −260°F then loaded into pressurized double-hulled vessels. (Picture a ship that's built like a gigantic thermos bottle.) In this state, natural gas occupies only 1/600th of the volume of its gaseous form. At the receiving terminals, the pressure is released (carefully.) and the gas returns to a gaseous state and is moved into pipeline systems like other natural gas.

---

[10] (Mouawad 2009)

[11] In the first half of 2009, gas sold for $3.50–$4.00/MCF at the wellhead. From the wellhead, gas is processed and transported via pipeline. Commercial end-users paid around $10/MCF for gas; while gas to residential users (who purchase far lower volumes) sold for $12–$13/MCF. (U.S. Energy Administration 2009)

Oil companies produce a great deal of natural gas, and natural gas companies produce a lot of oil. That's because oil and gas are typically found in geological association with each other. After all, the same natural earth processes "manufacture" them both. Indeed, coal is pretty much the same stuff too … just at a different age and stage of maturity. Although gas is measured in cubic feet, coals in tons, and oil in barrels, all three are carbon-based substances. (As are diamonds, interestingly enough.)

Nuclear Power

Of those countries that use nuclear energy to generate electricity, the U.S. uses far less than most. France generates roughly 75% of its electricity from nuclear plants.[12] Most European nations run on 25–50% nuclear power. Here in the U.S. less than 20% of our electric power comes from nuclear reactors.

The technical and scientific processes that go into developing and operating nuclear power facility are far beyond the scope of this book, and indeed the author's comprehension. But here's the basic process: A nuclear reactor core generates tremendous quantities of heat. In a power plant, water in sealed tubes circulates around and through the nuclear reactor, like the coolant in your car's radiator. This circulation does two things: it cools the reactor core, and boils the water into steam. The steam drives turbines, which are connected to electric generators.

As the steam expands inside a turbine, it cools and condenses back into water. The water is cooled, and then recirculated back into the nuclear core, where the cycle repeats. The only by-product of this process is warm/hot water (lots of it), but this water is circulated *separately* from the sealed core-heat-steam loop … so it is not radioactive, and poses no environmental threat once it cools.

---

[12] France also successfully reprocesses spent reactor fuel (which, for some reason, seems to still be a problem — if not an enigma — for the U.S.).

Nuclear power generation yields no emissions. No carbon output, no exhaust fumes, no undesirable effluent, except warm water.

There is only one "output" concern surrounding nuclear energy, and that's "spent fuel" (*i.e.*, radioactive material that lacks the potency needed in the reactor core, but is still radioactive and thus requires special handling/storage). The safe handling and storage of this sensitive material is the biggest issue facing the nuclear power industry in the U.S. Other nuclear power-generating nations reprocess their spent fuel and have successfully mitigated the problem of where and how to store it by those reprocessing methods. The U.S. has in place a policy (some 30 years old) against the reprocessing of spent nuclear fuel. As I understand these technologies, such material can quite practically be reused in a variety of applications, and the final residual matter is of such small quantity that it can indeed be stored in a relatively small space, entrained in a non-reactive form that is impervious to the elements. This would obviate any concerns about accidental release, discharge, or chain reaction.[13]

Most of us are familiar with (or have at least heard of) the "Manhattan Project," our nation's top-secret atomic bomb development project during World War II. I've recently heard a number of voices suggest that the U.S. should initiate an "Energy Manhattan Project" to solve our energy problems.[14] The idea would be to focus our considerable energies and brain power in an effort to build new reactors around the country.

That's a really good suggestion, in my view ... but it is fraught with fear, environmental concerns, and objections. Many of these

---

[13] Material stored in such a form cannot be reprocessed or synthesized into high-grade nuclear material, no matter how hard someone may try. This makes it useless to any group with ill intent who might wish to steal/hijack spent fuel and convert it to use as a weapon.

[14] I recently heard a caller on a radio talk show suggest that very thing during a discussion of whether we need a national initiative in support of nuclear energy could happen in today's social and economic climate. The caller (himself a retired nuclear scientist), said simply, "We already did that in the 1940's ... it was called the 'Manhattan Project' ... let's start building some nuclear power plants!"

concerns hearken back to the 1986 Chernobyl disaster in the Soviet Union, or the 1979 Three Mile Island incident in Pennsylvania. At Three Mile Island, one of the plant's two nuclear reactor cores (Unit 2) suffered a partial meltdown ... the progress of which was halted by automatic safety systems.

THREE MILE ISLAND – NOTEWORTHY FACTS

Here are a few interesting (yet little-known) facts about the Three Mile Island Nuclear Generating Station near Harrisburg, PA:

First, while the Three Mile Island accident was certainly serious, the Nuclear Regulatory Commission (NRC) notes that the accident resulted in no deaths or injuries to plant workers, or to members of nearby communities.

Second, and perhaps more surprising (to non-residents of the area): the Three Mile Island power station continues to operate today, generating an average of 6,800 Gigawatt-hours (GW-h) of electrical power annually. Although the entire plant was shut down after the accident in Unit 2, the undamaged Unit 1 was restarted and brought back online in 1985. Since then, Unit 1 has operated continuously (except for routine maintenance outages) for nearly 25 years.

The fact is, nuclear power is extremely safe, and has been doing very well here at home ... albeit on a smaller scale than perhaps it should be. Here are a few key facts, of which many Americans aren't aware:

- In 2007 the U.S. generated about 20% of our total electrical power needs - some 806 million Megawatt-hours (MW-h) - from nuclear energy.

- Since the late 1980s, U.S. nuclear power plants have provided over 500 million Megawatt-hours of clean, reliable power every year.

- There has never been a radiation-related fatality in any American nuclear power plant; and no one has suffered any radiation-related injury anywhere as a result of nuclear power-generation activity in the U.S.[15]

---

[15] We're talking here about "nuclear-power-specific" causes, such as radiation exposure, explosion, etc., not electrical *injuries*. Ever since the first time Benjamin Franklin's wig

(-more-)

In terms of our needed future electrical generating capacity, nuclear power seems to be the one near-term major source of viable and long term potential. The good news in the nuclear arena is that there are currently some 13 applications in process for new power plants in the U.S. There are also technological advances in potential of developing small nuclear plants which can be built far underground, and scaled to fit such small applications as a single town's needs.

Inexplicably the currently proposed "Energy Policy" by the new administration seems *silent* in terms of further development of this important potential energy resource. Yet this is a non-carbon producing energy source ideally suited to electricity generation. There is undoubtedly a story behind this oversight/omission and one which bears closer scrutiny. (More on governmental policy below.)

Hydroelectric Power

A proven, effective, and clean source of electrical power generation, hydroelectric power (power produced by the movement of water through dams, waterfalls, and the like), still accounts for only about 2.5% of power generation in the U.S.

Expanding hydroelectric power generation seems a simple solution to some of our energy needs, but again it triggers concern among the environmental community for a variety of reasons. First, it requires the building of dams to hold water under controlled flow conditions, upsetting the natural flow of drainage waters. This in turn impacts certain biological species of fish, other

---

smoldered after a brush with lightning, people have managed to injure themselves (and others) through electrical mishaps large and small.

My point is about the safety of nuclear power-production technology, not the absence of any risk from electrical output. (In other words: While nuclear power generation has a proven safety record, the electric power it generates poses the same risk as electricity produced any other way. So if you stick a fork into the toaster at breakfast, you're on your own.)

water-dependent wildlife, algae, etc. It also affects water usage and ownership rights of both upstream and downstream interests.

Water rights, and the laws pertaining thereto, are of themselves matters of enormous magnitude and a practice of law in and of itself. When you add in the legal ramifications (and capital intensive nature) of dam-building, it is almost unimaginable that a massive electrical generation project — including the construction of 29 separate hydroelectric dams, in seven states, along the Tennessee River — could ever have been undertaken or completed. Yet we did just that, in the 1920s. The TVA, or Tennessee Valley Authority, was a massive public works project that came on-stream in 1933. Today, the TVA is the single largest supplier of hydroelectric power in the U.S.

The TVA's present generating system also utilizes coal fired plants (which actually provide about two thirds of all TVA's power output). Nuclear plants and wind turbines are also used in its operations as the largest single "public" power provider in the U.S. Hydroelectric power only accounts for about 10% of TVA's total power output.

We must note that the TVA project began in the early 1930s, when the United States (and indeed much of the world) was in the throes of a Great Depression. In those days, the environmental lobby and our overall consideration of and concern for the environment did not represent anything like today's level of interest, consternation, or litigation.

It is highly unlikely, in my opinion, that we will be building many (if any) more hydroelectric power dams in the U.S. for the foreseeable future. Indeed, over the past decade many more dams have been removed than have been built.

INDUSTRY SIZE AND SCOPE
Just How Much Fuel Do We Use in This Country?

The following figures on U.S. energy production and imports come from the Energy Information Administration's "Annual Energy Outlook" for 2007 (the last full calendar year for which data are available). These numbers show U.S. Production and Imports, which we add together to derive a "total consumption" amount.

The United States is a net importer of energy, meaning that we consume virtually *all* the energy we produce and still have to import more to meet our needs. It's also understood in the energy industry that, for most part, we *consume* all the electrical power we produce. We, as a nation, don't really store or preserve electricity; it hits "the grid" the instant it's generated. This axiom — that we produce only as much power as we need, and thus import only that fuel needed to produce it — is what ties production figures to consumption.

So, in order to arrive at an understanding of how much energy we *use* in the U.S., start with the total that we *produce* domestically, then add to that the energy we *import* each year. The total is our annual consumption. Here it is in equation form:

PRODUCTION + IMPORTS = TOTAL ENERGY CONSUMPTION

In North America, we typically measure the relative heat value of fuels produced (or consumed) using the British Thermal Unit, or

"BTU."[16] These measurements are expressed as quadrillions of BTUs ("QBTU" or "Quads") per year. An engineer or scientist may have some idea of what quadrillion BTUs would look like, but most of us certainly don't. So here's one way to think of it: A "quadrillion" is one million-billion, or 1,000,000,000,000,000. (For reference: that's about five hundred times the number of ants on the planet. In other words: a *lot*.)

Table 1: U.S. Energy Production + Imports (QBTU, 2007)

POWER SOURCE	QBTU	% of U.S. Production	% of US Energy Needs
Crude oil & associated liquids	10.73	14.8%	10.2%
Natural gas & gas plant liquids	22.25	30.7%	21.2%
Coal	23.50	32.4%	22.4%
Nuclear	8.41	11.6%	8.0%
Hydroelectric	2.46	3.4%	2.3%
Biomass	3.23	4.5%	3.1%
Other renewables (wind, solar, landfill methane)	0.97	1.3%	0.9%
Other (incl. solid waste, hydrogen, methanol)	0.97	1.3%	0.9%
TOTAL DOMESTIC ENERGY PRODUCTION	**75.52**	**100.0%**	**72.0%**

Net Imports[17]	29.42		28.0%
TOTAL ANNUAL U.S. ENERGY NEEDS	**104.94**		**100.0%**

Table 1 (above) shows the numbers for 2007. Figure 1 presents a visual breakdown of the Domestic Production numbers of Table 1, excluding imports.

---

[16] The term "BTU" is used to describe the heat value (energy content) of fuels. BTU has also become a common reference unit for describing the power of heating and cooling systems, such as furnaces, stoves, barbecue grills, and air conditioners. When used as a unit of *power*, the technical term is actually "BTUs per hour" (BTU/h), though this is often confusingly abbreviated to just "BTU."

[17] Two-thirds of this quantity — roughly 20 Quadrillion BTUs — is crude oil.

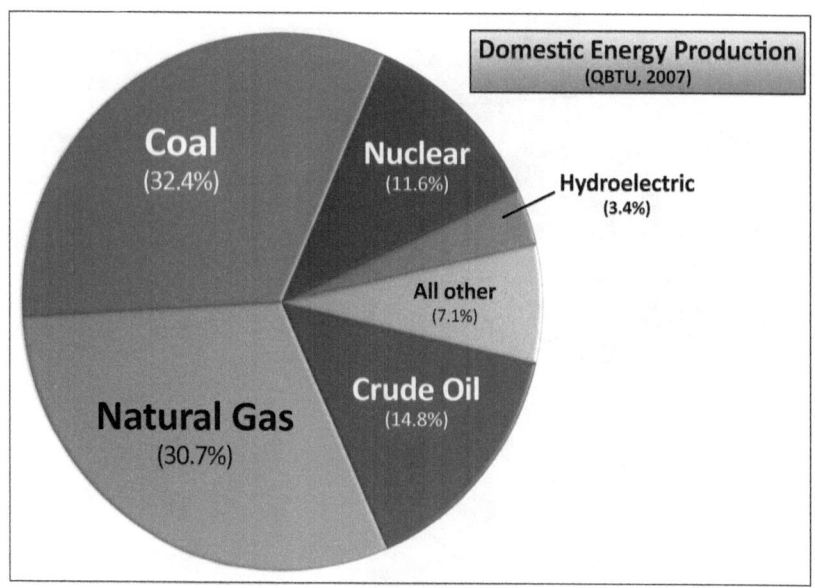

Figure 1

Table 1 and Figure 1 show our domestic energy *production*. In contrast, Table 2 and Figure 2 (below) focus on annual U.S. energy *consumption*.

Table 2: U.S. Energy Consumption by Source, 2007

POWER SOURCE	% of Total Consumption
Petroleum & liquid fuels	39.98%
Natural gas	23.25%
Coal	22.31%
SUBTOTAL: OIL, GAS, and COAL	85.54%
Nuclear	8.25%
Hydroelectric	2.41%
Biomass	2.60%
Other renewable sources (wind, solar, landfill methane)	0.01%
Other (incl. solid waste, hydrogen, methanol)	< 0.002%
SUBTOTAL: All other sources	13.27%
TOTAL ANNUAL ENERGY CONSUMPTION	≈ 98.81%

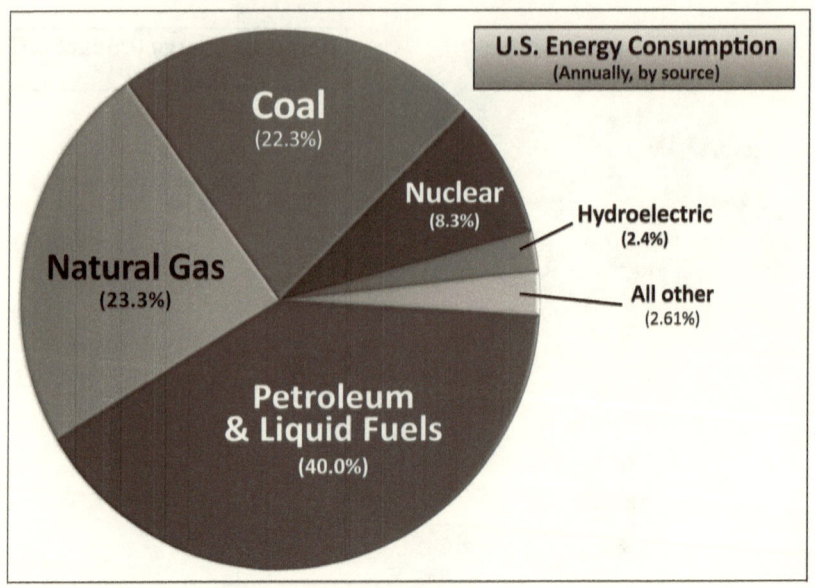

Figure 2

For our purposes, the significance of these figures is not the quadrillions of BTUs produced or consumed, but rather the relative *percentages* of our energy supply derive from which sources. Oil and natural gas combined represent over 60% of the energy we consume in the U.S. If you add in coal, the "big three" provide 75%. Renewable sources provide less than 3% of our energy.

Table 3: U.S. Energy Source Production Surplus / (Deficit), 2007/2008

POWER SOURCE	U.S. Production	U.S. Consumption	Surplus/ (Deficit)
Crude oil & associated liquids (million bbls/day)	4.95	(19.48)	(14.53)
Natural gas & gas liquids (billion cubic ft/day)	63.58	(63.55)	0.03
Coal [2007] (million tons)	1,146.6	(1,129.28)	17.32

Table 3 provides a side-by-side comparison of what we produce, versus what we consume, by source. Table 4 compares electric generation by power source.[18]

Note the disparity in U.S. supply and consumption of petroleum and liquid fuels. That source by itself represents 40% of our nation's energy consumption, and is the one at the greatest risk of interdiction from boycott, sabotage, and/or price fluctuation. We are effectively self-sufficient in natural gas and coal; yet we can produce only 25% of the oil we use. *That means we buy and import the 75% difference.* Even at low crude oil prices ($50/bbl) and today's reduced levels of consumption, *the U.S. is still spending $450,000 a minute on imported oil.*

Table 4: U.S. Electricity Generation by Power Source

FUEL SOURCE	Billion kWh/day	% of Total Generated
Coal	5.40	49.32%
Natural gas & other gases	2.18	19.91%
Crude oil/fuels	0.23	2.10%
SUBTOTAL: OIL, GAS, and COAL	7.81	71.32%
Nuclear power	2.20	20.09%
Hydroelectric power	0.70	6.39%
Geothermal	0.04	0.37%
Wind, solar	0.13	1.19%
Other (wood, waste, misc.)	0.07	0.64%
SUBTOTAL: All other sources	3.14	28.68%
TOTAL ELECTRICITY GENERATION	10.95	100.00%

[18] The figures in these tables are also from the IEA, and are from 2007 and/or 2008 depending on availability of data. (Some "other" categories are combined to provide clearer comparisons.)

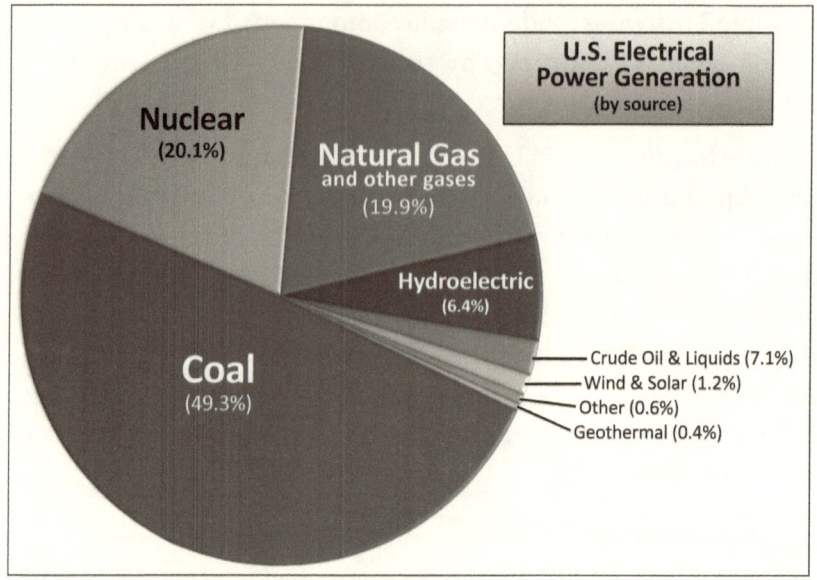

Figure 3

Just How Big is This Whole Energy Business?

It's *huge.* The oil and gas industry (including exploration, production, refining, and distribution) directly employs almost 400,000 workers in the United States. Coal, with its mining and power generation operations, employs another 150,000 or so. Nuclear and hydroelectric power generation accounts for 50,000 jobs. Power distribution activities — building and maintaining power lines, reading electric meters, in communities large and small — engage yet another 425,000 of us.

All told, the U.S. energy industry represents just over one million significant jobs, annual payrolls approaching $100 billion, and annual revenues of over two *trillion* dollars. (Not to mention massive annual capital budget outlays to build and maintain the structures and infrastructures supporting our use of power. These outlays likely represent another 300,000 to 500,000 good jobs ... all right here at home.) According to the American Petroleum Institute (API), the U.S. oil and natural gas industry supports - directly and indirectly - more than 9 million American jobs. (API Newsroom 2009)

Perhaps most important of all, our energy industry makes all the things we use work for us.

THE CURRENT ECONOMIC ENVIRONMENT

In the year since I undertook this project, much has changed in the world's economies, the consumption of fuels, and energy demand. Recognizing that yet more will change by the time many read this, I must nevertheless focus on a set of circumstances at a particular point in time in order analyze and comment upon the state of the industry in relation to the economy.

After several years of extraordinarily good economic times, the economies of the United States and the world in general have declined rapidly, due principally to the burst of the U.S. "housing bubble" in mid 2007. The causes, the blame, and the recriminations for that event — and other economic factors — will doubtless be assessed for many years into the future. But for now, all countries and their governments must focus on "righting the economic ship" in order to move forward.

That is the challenge facing Barack Obama, the recently-elected American President. Then-Senator Obama ran and won his campaign largely on issues surrounding the failing economy. He now has the unenviable task of fixing many problems. While we wish him well, for our own sakes as well as his, he is relatively inexperienced in the world of finance. (To his credit, it appears the President has assembled a formidable team of economic advisors to assist in this massive endeavor.)

In this same vein, the new President truly represents a sea change in American politics. He is not only the first U.S. President of color, but his party has significant, if not dominant majority representation in both Houses of Congress. That may prove to be positive for the country; it may not. Time will tell.

America's Auto Industry

With our present economic crisis, another industry, (often aligned with energy and, most assuredly, mutually dependent upon both it and the economy) is having more than its share of problems: the automobile industry in the U.S. The "credit crunch" has precluded many prospective buyers from being able to purchase cars, and the vicious cycle tightens. The manufacturers themselves have been forced to stiffen their terms for financing auto purchases, while facing tighter credit demands by their parts suppliers. The result is another industry that's in crisis and looking for another government bailout…they've already had significant funding from the government, now they're going back for more.

All three traditional U.S. manufacturers (Ford, GM, and Chrysler) are in serious trouble. GM and Chrysler filed for bankruptcy in 2009; both have since emerged from bankruptcy but their long-term viability remains to be seen. After decades of ignoring the threat (posed largely by Asian manufacturers) of smaller, more fuel efficient, and higher quality automobiles, perhaps some or all of those U.S. firms should indeed be allowed to fail. There would be however, serious ramifications in such a determination.

The "big three" car makers — especially General Motors — have billions (if not trillions) of dollars in "Commercial Paper" (bonds, notes, etc.) spread throughout the world's investment portfolios. A default on that paper would send further shock waves through the economies of many other nations and institutions at a time when the entire world's economy is in a fragile state.

IF THE EARTH STOOD STILL
AN UNPLEASANT SCENARIO

Think on this (hypothetical) scenario for just a moment:

You wake early one morning and reach for the switch on your bedside lamp. It doesn't come on. Muttering about a burned-out bulb, you get out of bed and flip on the light switch on the wall ... but still no light comes on. You want to check the time, but your clock radio is dark. Great, the power is out.

We've all been through power outages at one time or another ... no big deal, right? But what if it's out across the whole neighborhood ... the whole city ... everywhere?

You need some food, but your car is out of gas. You could walk to the supermarket (if you live in a city, as most of us do), but they're not open because there would be no lights or refrigeration. (Nor would they have received any shipments of groceries from their suppliers, because they are out of fuel too.) And even if the store were open (without lights or refrigeration), their cash registers won't work. Some vendors might open for business, but you can bet they'll take cash only ... credit card transactions don't work without power.

While you're in town (on foot), you think about getting some cash to have in your pocket till the lights come back on. But the ATM isn't working, and the bank is closed as well.

If you live in a large city, the trains and subways will be shut down. The airports too. Hospitals can keep core systems running on a generator for a time, but most medical facilities (even emergency rooms) are shut down except for catastrophic needs.

You want to contact some family or friends, but your home phone is dead. And your cell phone battery is low, although it wouldn't matter if it were fully charged ... the cellular/mobile networks are down too. (No power, remember?)

Years ago, you stopped keeping paper records (or even prints of your family photos), preferring instead to keep them in the clutter-free world of the "online cloud" at Yahoo or Google or AOL. That includes all your phone numbers and addresses, your bank statements, tax documents, health records, recipes ... even snapshots of children or grandchildren. Those may or may not still be safe on a server someplace, but you can't get to them ... at least not now. You can't look anything up, nor can you phone anyone else who can.

The food supply is shut down as well: Dairy farmers cannot timely milk their cows without power milking machines so their herds suffer illness and lost production. (Ditto large-scale livestock operations — chickens and hogs in particular — that depend on automatic feed/water systems for their animals' survival.) Crops wither without powered irrigation systems; even if the crops are ready for harvest, farmers can't bring in the crops without their big machines. Fruits and vegetables rot on the vine or in the orchard.

Now for the really bad part: Suppose it doesn't get any better tomorrow, either. Or the day after. <u>Now what</u>?

Is such a scenario likely? **Probably not.**

COULD such a thing happen? **Of course it could.**

Energy — electricity and combustible fuels, mostly — makes things work. We couldn't sustain our modern lives without it.

There are numerous sources of energy, each with its advantages and negative features. In the prior chapter we looked at those sources. In this section we'll look more closely at their relative benefits, and how we utilize them. They are, in brief, used for three primary purposes: (a) to generate electrical power; (b) for transportation; and (c) for direct production of heat (for industrial processes, for home heating, and for cooking). These are the things that we need to work for us ... *every day*. And without which we'd be utterly lost.

To be sure, there was a time — still within memory for some — when human life and civilization grew and thrived without electric power, or transportation fueled by anything other than hay and oats.[19] But with the Industrial Revolution, humans progressed from agrarian to industrial economies. We replaced manpower and horsepower with machine power ... machines fueled by coal, oil, and gas.

The U.S. doesn't have a monopoly on energy use, but we do use an inordinate share of the world's available supply. Is that because we're bad guys? No, it's because we have the most successful and largest economy in the world; we have a lifestyle and level of comfortable living which is the envy of the rest of the world's populations. They may hate us or admire us, but they all know we've got more than most ... and many hundreds of thousands aspire to live and work here — legally or illegally — rather than stay home.

Almost everyone in the rest of the world aspires to have many of the things we have, and many of those people and their governments are moving rapidly to catch or surpass us. Two of the largest such nations (China and India) are making huge strides in that very challenge. And a key component of that growth effort is their increasing usage of electrical power. They will, therefore, require a great deal more generating capacity than they now have.

---

[19] Certainly, some cultures on our planet still live this way ... and it is they who would likely be best-equipped for survival if, one day, electric power were to suddenly just vanish.
Indeed, such a civilization would be entirely unaffected by the cessation of power supplies ... because they don't use them.

WHAT WE DO WITH ALL THAT OIL, GAS, and COAL

Every day, we refine some 20 million barrels of crude into some 400 million gallons of gasoline, jet fuel, motor oil, and other petroleum products. We drive our cars and fly fleets of aircraft, commercial and private alike.

While we import a great deal of *fuel* into the U.S., we generate plenty of electric power ... billions of watts of it. Electricity is measured in watts, kilowatts (1,000 watts) and Megawatts (1,000 kilowatts = 1 million watts). Electric power is generated by turbines in plants powered by the various fuels previously discussed: oil, natural gas, coal, nuclear power, and water (hydroelectric dams where water pressure drives the generating turbines). We use all that electricity to heat and cool our homes and offices, power our factories and run our businesses. Alternative sources and methods of electrical power generation are now being seriously developed and applied; solar, wind-driven turbines, and others.

In 2006, the electricity generated in the U.S. was just over 4 billion Megawatt hours (MWh). The breakdown percentages of power generation (by fuel source) were, for that year: coal 50%, natural gas 20%, nuclear 20%, hydroelectric 5%, renewable fuels and "others" 2.5% (including wind and solar).

The consumer price sales value of that U.S. electricity (2006) totaled over $325 billion. That's over *three hundred billion dollars*, yet even that number is only about half our crude import bill for a year, at $150/barrel (as it has been recently, and likely will be again.)

[Just as a point of reference, in the part of the country where I live the average residence uses about 650 kilowatt hours of electrical power per month.]

AN INCREDIBLE 20th (ENERGY) CENTURY

My father and mother were both born in Indian Territory (as the state of Oklahoma was known, before it achieved statehood in 1907). They were born in 1896 and 1902 respectively; into a country of 45 states, and lived to see amazing innovation and change, as well as the addition of five more states — the last 10% of the nation — to our Republic. Their lives spanned an era of horseback transportation, with no paved roads, to the interstate highway system; from the telegraph for long-distance communication, to telephones, radio, television, and even faxes.

The 20th Century saw the dawn of the age of flight and a man on the moon; the invention of not just the airplane, but also the great airships, and the helicopter as well. From candlelight and gas lamps, to the blazing lights of Broadway, Las Vegas, and all the world's major cities. Not to mention nighttime stadium sports, including baseball, football, soccer, softball … from Little League to the Big Leagues … all made possible by electric lighting.

In that period, we went from using the gas from coal and coal-oil to drilling oil and gas wells in thousands of feet of water and many thousands of feet beyond into the ocean floor. Nuclear energy was originally harnessed for use as a weapon in WWII, but moved thereafter to power generation.

Air conditioning, the vacuum cleaner, the electric washer/dryer, dishwashers, garage door openers ... all were similar (energy consuming) 20th Century innovations.

From the Kodak "Brownie" camera emerged candid/family snapshots, first in black-and-white and later in color. Mr. Edison's movie camera gave us silent films, which gave ways to "talkies" and Technicolor. Those things led us to today's video cameras and digital cameras. The hand-cranked, centrally-switched local telephone became today's wireless phones and PDAs.

The cash register became the electronic card scanner; adding machines and calculators became the computer (undoubtedly the single most significant invention of the latter 20th Century). Having emerged as building-sized mainframe behemoth systems (with performance that pales in comparison to today's laptops), the computer's footprint shrank to handheld size, while its power and capabilities grew exponentially.

With the final decades of the century came the worldwide web, which — like the computer — represents yet another paradigm shift in the way the world works and plays, and the way we communicate with each other.

What a 100 years that was. (And as icing on the cake: the 20th Century did not end with a total collapse of our technical infra-structure — as many believed it might — due to large-scale Y2K-related system failures.) While credit for this staggering progress goes to many wonderful discoveries, the product of much hard work by many very smart people worldwide, it was all made possible, in one way or another, by abundant supplies of energy.

We've become more productive and more energy efficient as the years have passed; we use much less energy per dollar of GDP than was the case in the mid 1900s. This is due, in part, to the decline of heavy manufacturing in the U.S., and the increase in more energy-efficient (or "greener") technologies, processes, jobs, and indus-tries.

POWER IS POWER
In Today's World (and Tomorrow's)

David Boren is a former Governor of, and U.S. Senator for, the state of Oklahoma. Since 1994, Boren has served as president of the University of Oklahoma. In his book, *A Letter to America*, Boren posits, "How much longer do we believe the U.S. will remain the world's dominant power?"[20]

To many of us Americans, the answer might well be, "I've never thought about it *not* being dominant," or, more simply: "Always." In fact, neither is likely to be the case.

We're Not Alone on the List of Major Powers

To most people, the term "power" (in the global sense) refers to a state's military might and/or economic strength. And while our nation is far from perfect, we certainly top the list — for now, anyway — of global powers using those criteria.

But the United States is only one of many major world powers. Several others are growing at unprecedented rates and may well

---

[20] (Boren 2008)

replace us as the world's "800-pound gorilla." Here's a brief look at a few of the bigger players with a significant role in the future of global power:

CHINA. With a population some five times that of the United States, a burgeoning economy, and a considerable natural resource base, China will likely be the first to challenge — or even surpass — the U.S., on both the economic and military fronts. (This is not to suggest the U.S. is headed toward military confrontation with China; but we must nonetheless pay attention to China's capacity, and its oft-stated desire to become a world leader in both arenas.) Already a nuclear power, China will not likely back down from any arms race.

Significantly, China – currently the U.S.' largest lender - is signaling its concern over our present financial policies.

INDIA. Similarly, India — with a population roughly equal to China's — has proven its prowess in the world of computer and software technologies. Its rate of economic growth over the past decade has been astounding. Though its economy currently lags behind China's, India is rapidly becoming an economic force to reckon with. One can only speculate as to India's military aspirations, but India is itself a nuclear power. India has a long-running territorial dispute with Pakistan, its fractious (and nuclear-armed) neighbor to the north. It appears likely that India too will focus on military strength.

JAPAN. Once a dominant military power on the world stage, Japan currently occupies the number-three position in terms of world economies, and will no doubt strive to maintain its economic status. While their economy seems relatively stable, it's unlikely that their economy will surpass America's in the foreseeable future. From a military standpoint, Japan maintains purely defensive forces. (It is precluded from becoming a global military power under the terms of the surrender accords signed between Japan and the Allied Powers at the end of World War II.)

RUSSIA. With its massive natural resource base and history as a global power, Russia would seem another likely challenger to the U.S. for both economic and military superiority. Its major problem is the "hangover" from the days of communist rule, and scars from the Cold War. Political strife and corruption continue to hamper economic progress. To one who is neither an economist nor a sociologist, it appears that Russia's attempt to shift directly from Communism to a capitalist economy was too large a "leap" to accomplish all at once. (This is unsurprising, given Russia's tumultuous political history: Centuries of Czarist rule were followed by revolution and Marxist Communism, Stalin's virtual dictatorship, and a forty-year Cold War.) Russia's notoriously-inefficient bureaucracies, together with its lack of experience with capitalist systems, have all but stalled that transition. In short, things fell apart, and the vacuum seems to be filled by another rising "strong-man" government. Time will show us how strongly Russia plays on the world stage as an economic and continuing military superpower.

THE EUROPEAN UNION. The EU represents the collective interests of twenty-seven nations, most in western Europe.[21] It includes long-standing friends (and some erstwhile enemies) such as France, Germany, Italy, and the U.K. Considered as a single economy, the EU generated an estimated nominal gross domestic product (GDP) that's about a half-trillion dollars larger than ours, amounting to over 22% of the world's total economic output. This makes

---

[21] As of late 2009, the European Union comprises twenty-seven sovereign Member States: Austria, Belgium, Bulgaria, Cyprus, the Czech Republic, Denmark, Estonia, Finland, France, Germany, Greece, Hungary, Ireland, Italy, Latvia, Lithuania, Luxembourg, Malta, the Netherlands, Poland, Portugal, Romania, Slovakia, Slovenia, Spain, Sweden, and the United Kingdom. (European Commission, 2009. Web page *"EU at a Glance"*; http://europa.eu/abc/european_countries/index_en.htm) (European Commission, The 2009)

the EU the largest economy in the world (ranked by GDP).[22] The EU is also the largest exporter and importer of goods and services, and the biggest trading partner to several large countries such as India and China.[23] The EU does not maintain a significant military force, but many of its member nations (including several nuclear powers) do supply military forces to participate in various EU peacekeeping and defense activities.[24]

There's More Than One Kind of Power

In reviewing the above summaries (which only touch upon a handful of economic powers), it's easy to get the idea that for a nation to have significant "power," it must have (a) a huge economy; and (b) considerable military might. Certainly, these are measures of "power" in the classical sense. But before we conclude that money and guns are all it takes for a nation to be powerful, we have to consider one other major factor, which is one of the key points of this book:

> *Any nation that depends on others for its energy needs is at great risk of sudden economic hardship (or worse), should its energy supply be interrupted.*

And that, dear reader, is the position in which our nation finds itself today (we've been here for decades). The fact is that we, the people of United States of America, face a very real risk to our energy supply: The threat of a potential boycott, embargo, or even blockade of our crude oil supply.

---

[22] (Central Intelligence Agency, 2009. Web page *"CIA — The World Factbook"*, https://www.cia.gov/library/publications/the-world-factbook/index.html)

[23] (Central Intelligence Agency 2009)

[24] A number of recent treaty agreements provide for greater organization in this arena, but the charter — designed to protect the sovereignty of its member nations, prohibits them from dedicating their armed forces to the control of one blanket military authority on any permanent basis.

Lest we forget: *it's happened before.* In 1973 OPEC embargoed oil supplies to any of the "friends of Israel" during the Yom Kippur War. I well remember the blocks-long lines at service stations. My children were just starting to drive then, and they wanted to burn a lot of gasoline. (In those days I was commuting 42 miles — each way — to my office in Houston.)

A decade later, the price of crude oil dropped precipitously, sending the U.S. oil industry into a near-calamitous tailspin. That drop in price was precipitated by political events in Middle East: the overthrow of the Shah of Iran; overproduction by some OPEC members; and temporarily-reduced consumption in the U.S. (and elsewhere in the world).

Recently, Russia — the principal supplier of natural gas to Europe — literally *cut off* supplies of gas to its European and Balkan customers over a purported a price dispute with Ukraine. (More likely the price issue was a convenient cover story for a Russian demonstration of political clout, allowing it to serve notice of its regional dominance upon its neighbors and energy-dependent customers.) That action was very effectively taken in the midst of an extremely cold winter in the region.[25]

---

[25] Interesting side note: Recently, Russia's crude oil output has actually *surpassed* that of Saudi Arabia, albeit only intermittently.

Robert Harston

ENVIRONMENTAL CONCERNS

A detailed, scientific analysis of how mankind and its production/consumption of energy affect Planet Earth is far beyond the scope of this book. Nevertheless, no discussion of energy is complete without touching briefly upon the environment and (more to the point) the influence of environmental coverage in the media on the energy debate. The aim of this chapter — like this book — is to address some public policy issues and how they apply to the broader topic of America's energy.

Mankind's Footprint

We humans, through our very presence on the planet, have a direct impact on the environment. There's no denying this simple fact. Yet the acceptance of this truism appears to be lost on some of those who engage in environmental debate.

The foregoing is not say we should ignore the environment, cut down every tree, or dump anything we want into the air or water ... nothing of the kind. Rather, the point is: "Anything humanity does will have some impact on the world around us. The trick -- as with everything in life -- is finding a balance. Yes, we citizens of the planet should do what we can to protect and preserve it, yet we must also be able to live here. And therein lies the rub: much of the environmental debate is carried on between groups that seem to advocate an "all-or-nothing" solution ... failing (refusing?) to see the need for balance between what we can do to improve the world around us, and what we must do to ensure the continuity of our society and our way of life.

Environmental Activism vs. Rational Thought

I respect our fellow citizens who are active in various environmental groups (some more so than others). The idea of working to clean up the air, water, and land is laudable and important. As a resident of Colorado, I appreciate the natural beauty of the Rocky Mountains, and I support (and benefit from) the efforts of those who work to find a healthy balance between productive land use and preserving the landscape.

I take no issue with those who work to save the rain forests, reduce the emission of pollutants into the air, preserve wetlands ... I can even understand the dedication of those who like to hug the trees. (It's the people who talk to the trees that I worry about ... but that's not really the point here.)

But there are those individuals and/or groups who, in their single-minded determination to win-at-all-costs, will completely overlook the environmental forest while embracing one specific tree.

Case in point: In 2008, Xcel Energy proposed to shut down two coal-fired power plants in Colorado. The plan was to replace these two plants with one gas-fired power plant, which would be constructed on the site of one of the existing coal plants after the latter was removed from service.[26] Not only does this repurposing of land avoid objections that a "new" plant might despoil pristine lands, it also enables Xcel to reuse much of the equipment, controls, buildings, etc. This is both economically and ecologically efficient, benefiting the environment and rate-payers equally. In other

---

[26] Natural gas-fired power plants are generally far more "green" than coal-fired plants. Carbon emissions from gas-fired plants are significantly lower than those from coal-fired plants, and far more efficient when compared BTU-for-BTU.

words, this plan represented a reasonable balance between the need for reliable power in the region and the desire to protect the landscape.

As I read this proposal, it seems to be an efficient, logical, and environmentally-positive plan. Yet this proposal is presently opposed by the local Solar-supporters, an environmental group which opposes the construction of ANY new fossil-fuel power plants, anywhere, under any circumstances. In other words, their stated policy is to dramatically reduce our power generation capacity, rather than see an existing facility replaced with a cleaner, more efficient one.

It is this type of extreme, strident, and short-sighted opposition that makes such organizations the type of environmental activist group that tries the patience of rational people, and which never seems to gain the trust or respect of the general public.

The Climate Change Debate

As discussed above, we generate electricity and transportation fuels from all presently viable sources. The problems are with our consumption habits, especially in the United States.

There's no denying that carbon-based fuels do create undesirable emissions. Recent improvements in scientific measurement, including space-based imaging and observation, have led us to this understanding. Yet the impact of those emissions on climatic conditions, and our planet's capacity to absorb and adapt to those changes, remains at considerable issue within the scientific community. Indeed, the very nature of most global changes (be they climatic, geologic, economic, or anthropological) makes the study of these phenomena largely retrospective ... in other words, we don't really know exactly what's happening until after it's happened.

This phenomenon is observable in most truly dynamic systems. Economists tell us that they cannot detect an economic recession until we're actually in the midst of one. Meteorologists can measure the wind speed of a developing hurricane, but they cannot predict the damage it might do on land until after it makes landfall (if it makes landfall at all). Similarly, while scientists can observe minute changes in the concentrations of greenhouse gases, and measure the carbon absorbed by the world's oceans (as increased acidity in the water), the long-term effects of these changes on the ever-changing environment remain to be seen.[27]

Fossil Fuels vs. "Green" Energy: Same Goals, Different Methods

Notwithstanding these open questions, the consensus among most groups — industry, environmental, and political — is that the reduction of carbon emissions is a worthy goal that will, at some unknown time, yield some unknown level of improvement(s) in the environment, at some unknown cost. (Everyone agrees it will help in some way or another ... but how much it will help, how long it will take, and how much it will cost is anyone's guess.)

In response to current public opinion, Congress has allocated huge sums to the development of "green" energy sources, the reduction of our "carbon footprint," and focused on alternative energies. With carbon emissions as the primary target, fossil fuels are right in the bull's eye.

Scientists agree that the single largest contributor to carbon levels in the atmosphere and the oceans is the combustion of fossil fuels in the form(s) of coal and natural gas (as industrial and power-generation fuels), and crude oil derivatives such as diesel, gasoline, and jet fuel. And, as discussed previously, our nation's present dependency on foreign oil sources is an inherently unstable, and terribly costly, state of affairs. These two factors clearly argue in favor of America reducing its consumption of fossil fuels

---

[27] For those interested in this topic, I'd suggest Ian Plimer's *Heaven And Earth: Global Warming - The Missing Science*, in which the author aggressively criticizes many of the current concepts and beliefs regarding climate change especially with regard to CO_2's affect on climate. His refutation comes from multiple scientific perspectives of astronomy, physics, geology, tectonics, climatology, oceanography, and archeology, among others. It's some 500 pages with over 2300 footnote references.

to a significant degree, in as short a time frame as possible. If, as, and when we do so, both the environment and our national security should be in much better shape.

But increasingly absent from these discussions is a healthy dose of pragmatism ... the understanding that wishing to eliminate the burning of carbon-based fuels is one thing; making it happen is quite another. We, as a nation, cannot suddenly sever all ties to fossil fuels. People have to get to work. Not "someday," *right now*, *every day*. Office buildings require lighting, elevators, and ventilation. Financial institutions and hospitals need electricity to function. Even the teams of research scientists developing next-generation alternative energies need to power their computers, download information from online libraries, and drive their gas-powered vehicles into remote areas to measure wind speeds or test new solar cells. And they need to do so today, not years from now after their work is complete.

Like it or not, our society depends upon reliable energy for power generation and transportation. At the present time, our infrastructure and our systems are all geared toward the use of coal, oil, and natural gas to supply those needs. Even as we continue the development of promising new technologies, we have to keep the lights on and the trains moving ... and that means the continued consumption of presently-available energy (which we hope one day to replace).

The production and distribution of energy across continents relies - as we've discussed before - upon massive infrastructure. Changing that infrastructure to accommodate different energies isn't like steering a new course in a speedboat ... it's more like changing course in an oil tanker: Sure, it can be done ... but no matter how far (or how quickly) you move the rudder this way or that, the ship can only change course gradually ... a process that takes a long time, and during which you'll travel a long distance.

This, then, is the overarching point of this chapter: to plead the case for rational thought and discussion in shaping energy policy and debate. Sure, we now understand (much better than we did a few decades ago) that our dependency on fossil fuels is an issue with implications that are not only political and economic, but also environmental.

Nevertheless, the migration of our economy, our industries, and our society from fossil fuels to alternative sources is exactly that ... a *migration*. It is a gradual, continuous shifting of production and consumption from one place to the next.

This migration will not happen overnight. And those who demand that it can, or must, simply fail to grasp the size of the challenge. Nor do they offer viable strategies to achieve it.

POLITICS
OPEC, THE PRICE OF OIL, and AMERICA'S ENERGY SHORTAGE

[NOTE: This section contains some opinions (mostly in the interpretation of facts and figures; e.g., "What does this mean to you and me?") I'll point those out, but the underlying numbers are factual.]

Texas energy tycoon T. Boone Pickens, along with some other smart, knowledgeable people, has been saying for a while that oil is going to $150 per barrel or more. Over the years, Mr. Pickens has made (and occasionally lost) fortunes by being mostly right (and occasionally wrong) about oil prices. He has experience and smarts, and so do many people who back his ideas. (More about Pickens and his alternative energy plans later in this section.)

Oil prices are a supply-and-demand issue (more demand, less supply). The world now produces about 83 million barrels of oil per day.[28] Demand is now down to below 80 million barrels due to the world's economic malaise. However; the citizens of China and India — the earth's two most populous countries — are only just now starting to drive automobiles on a significant scale.

---

[28] Only about 90% of that figure is actual *crude oil;* technically, the remaining 10% is not oil but is "associated liquids," which flow out of oil wells along with the crude itself. But for practical purposes, it's all considered "crude" in this context.

Additionally, these populations need (and consume) vastly more electric power for their burgeoning industries. That demand will likely accelerate dramatically as economic conditions improve in the next few years. In fact, China is now building one new coal-fired electrical power plant *per week.*

At the present, with the world's economies virtually all in recession, energy consumption is markedly decreased and the demand for coal, oil, and natural gas are significantly reduced. As a result, the prices of all those commodities are also materially lower than just a few months ago. But this condition is not likely to last, and we would be well-advised to develop alternative energies and to increase domestic supplies of oil and natural gas while we have this respite in demand, prices, and development costs.

The Gross World Product is expected to expand several times in the next few decades. Even with efforts to "go greener" the demand for energy will be greatly increased as this worldwide growth occurs. The world currently burns some 1,000 barrels of oil per second,[29] and we burn over a billion *tons* of coal a year in the U.S. alone.

Many factors contributed to 2008's very high oil prices ($150 barrel); chief among them was the relative weakness of the dollar at the time. (The dollar had been weak for a few years; since crude oil prices worldwide are standardized on the U.S. dollar, our currency's weakness may have accounted for 10 — 15% of that price increase.) There was also considerable "speculation" in the oil markets by commodity traders (as there is in all commodity markets) that may have accounted for another 20% to 30% price premium.

If those (my) assumptions are correct, the current world producer price of oil "should" probably be $50 — $100 per barrel. Yet all

---

[29] Tertzakian, Peter. *A Thousand Barrels a Second: The Coming Oil Break Point and the Challenges Facing an Energy Dependent World* (McGraw-Hill, 2006). (Author's note: This is a most interesting history of the world's fuel source progression, and an analysis of where we may be headed.) (Tertzakian 2006)

we know is that last year it was $150 a barrel and we were paying nearly $4 a gallon for gas at the pump. At the time there was no shortage of politicians who said "it's Big Oil's fault, so let's blame them [again]." Such knee-jerk finger-pointing ignores OPEC and the world supply/demand balance, and it dismisses the fact that we have not had any semblance of a cohesive energy policy in this country for over fifty years.

In terms of rapid change, in the past few months the price of gasoline has dropped about 70% as the price of crude has declined; and that certainly makes life easier, day to day. But I'm not betting it will stay down there for long. In fact, crude prices have essentially doubled from their lows of only a few months ago.

Just Who Is "Big Oil," Anyway?

The pejorative catch-all "Big Oil" is a tried-and-true synonym for the major oil companies in the U.S. This term has been around for as long as I can remember. It was routinely used to refer to the biggest oil companies of the day: such as the Standard Oil companies, Shell, Texaco, and others from the past. At some time in history, "Big Oil" was an appropriate moniker for these American firms to which it referred ... but no longer.

At one point in the early 20th Century, perhaps the most derisive of all oil company references was the long-lasting term "the Seven Sisters." These were primarily large U.S. companies which included most of the affiliated Standard Oil Companies, and a few others aggregated into a consortium in the 1920s.[30]

---

[30] The "Sisters" included Standard of New Jersey (now the Exxon of Exxon-Mobil), Standard of New York (SOCONY, now the Mobil of Exxon-Mobil), Standard of California (Calco, now Chevron), Anglo-Persian (now BP, also the parent of the former Standard of Indiana Pan Am/Amoco), Gulf Oil (now part of Chevron), the Texas Company (commonly known as Texaco, until its merger into Chevron), and Shell Oil.

Other references include Atlantic Refining (later merged with Richfield to become ARCO), as well as Mexican Oil, as original members of the Seven Sisters. Yet this doesn't seem logical; Atlantic was originally a refiner and minor exploration player, not a large-scale producer of oil. I know nothing about Mexican Oil. For certain, not even *I* am old enough to remember which they all were.

At least one source suggests that then-Secretary of Commerce Herbert Hoover (later our 31st President) was involved in, if not the driving force behind, the formation of that consortium. Our government (wisely) wanted to be sure the U.S. oil Industry was not excluded from then-rumored (and later proven) gigantic Middle East oil discoveries. Forming that group provided enough leverage, through control of U.S. distribution systems, to force the European oil Cartel to allow U.S. companies to become "players" in the Middle East oil development then just getting under way.

Since the 1960s or so, "Big Oil" doesn't apply for any U.S.-based or U.S.-operating firm (or group of firms). "Big Oil," in today's world, is now the government-owned and — controlled oil companies of the larger producing countries of the world. The "New Seven Sisters" are:

- Aramco (Saudi Arabia)
- GazProm (Russia)
- CNPC (China)
- NIOC (Iran)
- PDVSA (Venezuela)
- Petrobras (Brazil)
- Petronas (Malaysia)

A few other major government-owned firms include ADNOC (Abu Dhabi), the Kuwait Oil Company, and Pemex (Mexico). All of these national firms own the land, the subsurface rights, the wells, the pipelines, the refineries, the tankers, and the rights to all the hydrocarbons and other minerals in their respective nations.

These countries have — understandably — determined to control and develop their own natural resources. Largely a combination of monarchies and other strong centralized governments, they have by fiat and/or edict simply taken over. In some instances they have negotiated with outside firms operating within their boundaries to compensate those outside firms for their investments. In other instances, they have simply seized the foreign companies' operations, giving the employees a week or so to leave the country. It hasn't always been pretty or fair, but it's now a settled fact as to who's in charge in those nations. Given that our principal suppliers of imported oil are under the control of centralized governments of one sort or the other, their exports to us can and will be monitored, rationed and/or priced to reflect our government's agreement (and compliance) with their policies regarding world-order and relations.

What's Our Problem?

Okay, opinion time. My experience and my observations over decades in the energy industry lead me to the following conclusion:

America's oil price/supply problem is mostly <u>politics</u>.

Much of the remainder of this volume is based this thesis.

The Congress(es) and Administration(s) of the past 50 or so years have been joined by the current Congress in blaming everyone except themselves. They (Democrats and Republicans alike) have all talked about — or hidden behind — a so called energy "policy" which has in fact been a "non-policy" of *not* producing our domestic oil resources. And this policy signaled to the foreign producers, "We'll buy and use *your* oil while we save our own ... so go ahead and raise the price." *And boy, did they ever get the message.*

In the 1970s and 1980s, I spent more than a little time in Washington, DC, trying to talk to members of Congress and various Administration executives about the need for a meaningful energy policy in this country. During those efforts, which spanned two decades, I was effectively told: "We've given up on oil and gas in the U.S. ... why don't you go on home?" That was more than twenty-five years ago.

During the past 50 years of "non-policy," our leaders also decided that importing from overseas was such a good plan that we essentially told our U.S. oil and gas companies (big and small alike), "Since we don't really need you or your oil, don't explore here. Oil rigs are noisy and messy. We won't lease you more Federal land for exploration, and we won't let you spoil any of our ocean views with your offshore drilling rigs."

Some of the politicians, in their wisdom, also opted to listen to the environmentalists and determined that we shouldn't build nuclear plants or other new sources of power generation (more about that later). Congress, serial Administrations, and environmental interests have also helped kill progress on technologies to recover oil from shale.[31] Other politicians and administrators simply did not consider energy an issue worthy of their focus. The net effect: *Our government seemed to give up on our domestic energy industry.*

OPEC: Major Factor in Oil Prices Today

The two major forces in world oil price and supply today are: (a) the demand for petroleum, which, on a global basis, is neither centralized nor managed; and (b) OPEC, which is both.

The OPEC cartel was formed in the 1960s to control the production and therefore the price of oil. It took about ten years for OPEC to get its act together, but since then it has been, and remains, extremely effective.

OPEC is a shining example of the Power of Monopoly. Only one of the individual OPEC member nations (Saudi Arabia) is the largest producer of crude oil worldwide; most of the other member states are relatively minor players. Yet by cooperating with each other, pooling their production and their collective political wills, they have enough production (some 30 million barrels/day, roughly a third of the world's total), to control and manipulate crude oil prices worldwide.

As of this writing, there are twelve OPEC member countries.[32] With the exception of Venezuela and Ecuador, OPEC's membership is entirely Middle Eastern and African. (See Fig. 4)

---

[31] We have lots of that too, by the way.

[32] (OPEC Secretariat 2009)

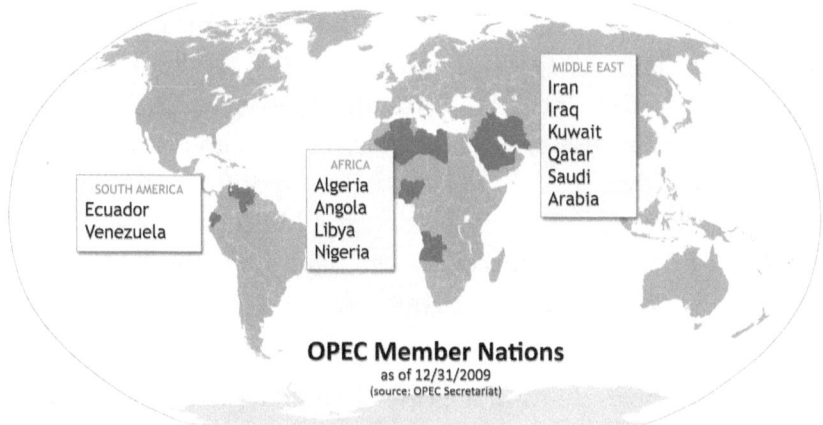

OPEC Member Nations
as of 12/31/2009
(source: OPEC Secretariat)

Figure 4

Today's single largest-producing nation is Saudi Arabia, which generates some 11 million barrels/day. Russia (which is not an OPEC member) is also a major producer of crude, at ≈10 million barrels/day. Although Russia isn't part of OPEC, our relationship with them just now is none too cozy.

When you review the history of world crude prices over the last half-century, the price of oil keeps edging upward until major consumers (like us) feel the pinch. This leads to increased pressure on our government to support the development of alternative fuels and technologies, then and we and other import-dependent nations start to make real progress developing alternative sources. Then, lo and behold: the price *drops*, crude makes financial sense again, and we return to our old habits.[33]

Do you think OPEC has a pretty good grasp of supply/demand and economics? *You can count on it.* Over the ten or so years I worked with (and got to know) representatives of various Middle

---

[33] A recent example of this phenomenon (on a smaller scale) occurred in 2008, when a sharp rise in domestic gasoline prices in early- to mid-year led to a spike in hybrid car sales, with corresponding drops of SUV and other lower-mileage vehicles. The media declared the Death of the Big Car, and it looked like fuel efficiency was the primary driver of car sales. But by autumn 2008, gasoline prices dropped by almost 50%. The story lost its momentum, and the relative sales of vehicle types (corrected for the economic downturn of that same period) leveled off to prior norms.

Eastern oil companies, I was always fascinated to discover how many engineers, managers, scientists and "Royals" held advanced degrees from the finest Western universities, in the U.S. and elsewhere. Education was, and still is, a very high priority within those circles. (Many of those I met held a specific combination of degrees: a Bachelor's Degree in science; a Masters Degree in engineering; and a Ph.D. in Economics. That's a formidable set of credentials, no matter where you're from or what you do for a living.

Anyone who underestimates the capabilities of energy industry leaders in the Middle East does so at their peril.

AN EVENING WITH OPEC

In 1980, while returning from a trip to the Middle East, I stopped overnight in London. Coincidentally, the editor of PI's international publications was also in London that day, and he had scheduled a late-afternoon interview with the Oil Minister of one of the largest OPEC producing countries.

The three of us met for aperitifs following the interview. The Sheik and I got on quite well, and carried on for an evening of conversation.

The Oil Minister of an OPEC nation is a person of considerable power. This particular Minister managed his country's policies for the production and marketing of their oil resources on the global markets. He represented his country at OPEC meetings, voting policies that would have direct impact on oil prices in America and elsewhere. Like so many of his contemporaries, the Minister held multiple college degrees — two geoscience degrees, plus a doctorate in economics — from Western universities. Needless to say, he was a highly intelligent, erudite gentleman of great social and political sophistication.

(A little background and context on OPEC: At the time of our meeting, OPEC had established oil production "quotas" (limits) for each of its member nations — as it often had, and still does today — in an effort to limit the volume of crude oil production across the cartel. [This is an age-old practice in commodities markets: manipulating the supply of a product to keep demand — and prices — artificially high.]

Over the decades, OPEC has occasionally imposed such quotas with varying degrees of success. This is because individual OPEC members — much to the chagrin of the cartel as a whole — are notorious for over-producing their individual quotas, to increase their own country's revenues. (In other words, the manipulators manipulate the other manipulators. That's irony, writ large.)

Most of the evening's conversation centered around the global petroleum markets and the industry generally. But during the course of the evening, I also experienced firsthand the remarkable ease with which a skilled diplomat can deflect direct questions and avoid giving direct answers. (We've all seen politicians practice that art in print or on television, but it was my first experience seeing and hearing it face-to-face ... and it was fascinating.)

I asked the Sheik directly, "Is [your country] over-producing its quota?" He replied, "No ... but if we were, it would be only to punish those who are." This non-answer ("No, but if so, it's someone else's fault") is a wonderful example of the diplomat's art. It's what lawyers call "alternative pleading" (I didn't do it. Or if I did, it wasn't my fault.)

Nevertheless, all in all, that meeting was an edifying experience with one of the true power brokers in the world oil markets at that time.

What's the Matter with the United States (and Our Elected Officials)?

We are funding both sides of the war on terror (even if that's not a politically-correct term, that's what it is) and we are running massive negative import/export balances. Meanwhile our elected representatives in Congress sit on their hands and toss out red herrings.

Are our representatives and officials all stupid, or ill-willed toward our future and prosperity? Of course not. I believe most (hopefully all ... but I sure have doubts about a few) are well-intentioned, hard-working, conscientious people. But they all have agendas ... their own, those of their constituents, and those of their geographic regions. Put many different agendas into any large project and strange things happen. Recall the old adage: "A camel is a horse designed by a committee." Politics indeed makes strange — and expensive — bedfellows.

Something Fishy

One red herring that shows up often in the media is an argument wrapped in a question:

> *"The oil companies say they want more places to drill here in the U.S., but they already have millions of acres under lease from the government they haven't drilled on yet. Why don't they drill there?"*

That's a valid question, one that's understandable coming from those who don't understand the industry. But here's the straightforward answer: *The industry has long since analyzed that acreage — through subsurface surveys geological studies — and they have already pursued drilling in areas where there was potential for significant oil or gas deposits. Most of the rest of the land shows no promise.*

Here's a logical follow-up question: "If that's true, why did they lease the acreage in the first place?"

Again, a reasonable question. Put simply, offshore tracts are made available in large acreage-lease offerings ... package deals, in other words. Oil companies have to bid by the *tract* ... they can't bid on just the few high-potential acres they want. They must lease the good with the bad, and they pay rentals on all of it.

This is a pretty simple concept, once you hear about it. But when such questions are tossed out and left unanswered (a common practice by cable news commentators across the political spectrum ... left, right, and center), they add to the confusion and mislead audiences. My point: Let's keep our facts straight and inclusive when we debate these issues.

Congress Takes Swift Non-Action

In the last Congress, the House passed a bill to permit additional offshore drilling. But in areas 50 to 100 miles offshore ... where most of the oil *isn't*. The Congress knows that, the oil industry knows that. But it gave Congress the opportunity to say, "We passed a bill, but you didn't drill." That phrase popped up numerous times during Congressional committee hearings last fall; it practically became a watchword of one anti-oil contingent (most of whom are Democrats, although certainly not all Democrats subscribe to their view.)

"Offshore" Has Changed Since 1970

Many people think offshore drilling means huge platforms, resembling floating oil refineries, sitting in the middle of Marina Del Rey or Puget Sound, spoiling the view, and belching smoke. But since its earliest days, offshore drilling — in most of the Western hemisphere, anyway — has been carefully controlled and monitored to protect the environment. From the North Sea to Gulf of Mexico, offshore rigs employ on-site oceanographers and scientists, with lots of technology at their disposal, to keep a close watch on every inch of the rig (and the drill stem). While there have been minor leakages on occasion, they are very few and very small. Offshore exploration, globally, is far cleaner and safer for the oceans than shipping.

For the benefit of the NIMBY ("not in my backyard") crowd, it's important to note that offshore drilling close to shore does not threaten the ocean view from that pricey Malibu retreat. An offshore drilling rig located just fifteen miles offshore *is invisible from shore* — not even the top of the mast extends over the horizon.

What's more — and this is a big surprise to most people — *you don't even have to drill offshore to find oil offshore.* Today's drilling technologies allow a land-based drilling rig to reach out horizontally as well as vertically. After drilling down a few thousand feet, they can steer the drill pipe laterally, extending the well out under the ocean floor — not underwater, but *under the sea floor* — to a distance of 30 to 40 miles into offshore hydrocarbon deposits. This methodology is cheaper than using an offshore rig, creates no visible disturbance above the ocean surface (or underwater), and poses almost no environmental risk.

But in order to drill into offshore deposits, even from a land-based rig, oil companies still need permission from whoever controls the area they drill into. Which means they still would need a lease from the government to drill into the near-offshore areas (where we already know oil and gas are) even if they're drilling from an isolated inland spot. And under the present political climate, that permission isn't likely to materialize any time soon.

ENERGY and the
CURRENT ADMINISTRATION

For all my commentary about the political malaise surrounding energy policy for the past six or so decades, most assuredly the current one is wasting no time or effort in attacking the problem. I applaud the effort and the attention.

What concerns me is not their desire to reduce our dependence on oil imports ... that's certainly what I want too. But sometimes a solution can be worse than the problem. The following assumptions are what concern me:

- "Global warming is caused by carbon emissions." It might be, or it might not. If this is a normal cyclical Solar event, we can't do anything at all about it.

- "Green Energy will create hundreds of thousands of new jobs." Maybe, or maybe it will only create "different" jobs, which will replace existing jobs which will be lost with no real gain and perhaps a net loss.

It is my belief that those who believe that climate change is, or can be, human-induced via carbon emissions are in fact emotionally invested in that position, not scientifically grounded. From my reading, it appears there is a paucity of solid science to support it. What I can find and have read, the earth's climate may well be changing as it has routinely done over millions of years ... it may be warming, it may be cooling. It's been much warmer and much cooler many times and species have evolved and developed and changed when that happens. Indeed, some have become extinct, as will others in the future … all as part of natural cycles.

Cap and Trade

The concept *du jour* among the environmental community, the Congress, and the current Administration is known as "Cap and Trade." The stated goal of Cap and Trade is to reduce carbon emissions by American industry and utility companies over time; the method proposes to permit different fuel users to decide how rapidly they can/should modify their processes to reduce their emissions. The concept employs a two-step process to restrict, over time, total industrial emissions of carbon dioxide and other green-house gases (also known as "GHGs").

STEP ONE is to impose a limit (the "Cap") on total emissions of GHGs by large-scale producers, such as heavy industry, utility companies, and the like. Each producer would obtain a permit under which it is allotted a quantity of "emission credits," allowing it to produce X quantity of GHGs/year. The quota for each would be tailored to the producer, based on the type of industry, the size of the entity, etc. Limits would start at one level, becoming more restrictive — allowing fewer emissions year-over-year — until the ultimate target limits are reached.

STEP TWO (the "Trade") allows different companies/industries to exchange emission "rights" under their various quotas. Thus, producers who can more easily or efficiently change their processes and reduce emissions may sell or trade any unused emission credits to other producers who may not be able to reduce their emissions as quickly. Thus, if one plant installs new emissions-control systems (or has a slack year), thus reducing its total production of GHGs, it might not need all its emissions credits for a given year. That user could sell its unused credits to a neighboring plant that might need additional credits for its greater carbon emissions.

Each producer of GHG emissions would determine its own me-thod(s) for reducing carbon emissions. The procedure for sequestering CO_2 emissions from the industrial combustion of coal and other carbon-based fuels is known as Carbon Capture and Sequestration (or "CCS"). CCS has been the focus of extensive studies, evaluations, and scientific papers generated on the subject. Development projects and test facilities abound.

The current cap and trade proposals for GHG emissions are similar to the cap and trade program enacted by the Clean Air Act of 1990, which reduced the sulfur emissions that cause acid rain (and, to the surprise of many, met those at a much lower cost than industry or government predicted).

Whether governmental allotments of emissions credits would be given away, sold, or some of each, is apparently still in question. It appears likely the federal government will *auction* these credits to industry. Either way, the government is already forecasting a revenue increase of between $600 and $700 billion for the next decade. So it certainly looks as if those permits will indeed be *sold* ... something for which we will all pay via increased prices for our energy.

Some call this a usage fee; I call it an "indirect tax." It would amount to between $1,000 and $2,000 per household in the U.S. per year for increased energy costs. Just a back-of-envelope-calculation tells me the electric bill for my home will go from about $125 per month to about $250 per month...but we must remember it's not a tax increase. Indeed, these costs will probably not all be loaded onto electricity rates, but they'll show up in our energy costs somewhere...we can all bet on that.

The March 2009 White House Agenda on Energy and the Environment promulgates five bullet points, including:

- Within ten years, save more oil than we currently import from the Middle East and Venezuela combined; and

- Ensure that 10% of our electricity comes from renewable sources by 2012, and 25% by 2025.

Other administrations have issued similar projections of 10% reduction by 2000, and 20% by 2020. Clearly we missed the first, and don't appear on-target to hit the second. My guess is we'll miss the new Administration's targets as well. Among other subset points in this White House paper are:

- Developing and employing clean coal technologies; and
- Prioritizing construction of the Alaska Natural Gas Pipeline. [Let's hope the government doesn't decide to "help" this project too much and delay its completion.]

They also promote the "responsible domestic production of oil and natural gas." Which sounds reasonable enough, depending upon your definition of "responsible." Of course, "promoting" a thing may mean no more than "talking about it frequently," which is something politicians do for a living. Or, it could mean "randomly enacting ill-advised and unworkable restrictions." Only time will tell.

Of particular concern to me (bordering on amazement) is that there is no mention of nuclear power development in the document ... a peculiar oversight/omission. In fact I submit that *anyone purporting to be a "green energy" advocate, yet who does not include nuclear power as a major component in their plan, is in fact not serious about their stated objective.*

A few years ago, the UK implemented a cap and trade tax to reduce emissions (and raise government revenues). I know of no one who believes that plan has effectively achieved either goal.

Recently the Environmental Protection Agency (EPA) has determined that CO_2 emissions are a substance that is harmful to the environment and as such it (the EPA) has the power to regulate and enforce action against those emissions. That represents a very effective first step in the "emission tax credits" program.

In southern California a program such as that proposed for the country has been in place for some period, its advocates claim remarkable success in reducing CO_2 and other targeted emissions. I have seen no figures on that project indicating its success or lack thereof.

FUTURE SOURCES of ENERGY
ALTERNATIVE FUELS and POWER SOURCES:
BIG NUMBERS and POLITICS

The energy business by its nature is enormous. Look at the numbers and volumes involved in the above paragraphs. Drilling for oil or gas is expensive; before you drill you invest a great deal of money determining the best place to drill (seismic surveys and other pre-drilling investigations and geological analyses). Building an oil refinery is a mega-million-dollar investment. Considering the purchase of a few oil tanker ships? Those run $50-100 million apiece.

Speaking of refineries, that's another shortfall we have in this country: There has not been a new refinery built in the U.S. for about 35 years. (The result of another set of environmental battles.)

You and a couple of neighbors aren't likely to pool your spare dollars and build a nuclear plant, or buy the acres of land and equipment for a solar-array generating system to serve just a few homes. In fact, putting up a wind powered generating turbine of any useful size (just one) can be a million-dollar investment.

These are projects for big investors and large companies (or government, if we want a truly inefficient operation). The energy businesses generate huge revenues and large dollar volumes, but in terms of return on investment, they are like most other business, just larger in terms of the number of zeros attached to their costs, revenues, and investments. Their percentage of profit returns are essentially the same as other businesses.

Here's an example of oil exploration costs (and risks) versus return: In the 1970's, one major oil company decided to drill a single "wildcat" well (a well in an untested, unproven area) in the Gulf of Alaska. After paying the U.S. millions of dollars for the concession leases, and investing more for seismic, geological and other subsurface evaluation processes, the company opted to drill one well to test the apparent high-potential geologic formations there. The total cost of that well, including the lease and analysis expense, was over $180 million. Adjusted for inflation, that's over $700 million dollars. Almost certainly the most expensive dry hole ever.

But the owners of that well did not invest their money out of idle curiosity. They felt they had to test those formations, and while they had reason to *believe* there were reserves underground, they had to find out for sure. That particular hole turned out to be a bad bet, but at that time a significant portion of scientific consensus was that there were large-potential deposits of oil and gas under the Gulf of Alaska (as indeed there are, about 1,000 miles farther north). But even with today's advanced oil and gas exploration technologies, we can't "see" 20,000 feet down into the ground. As it turned out; those "source beds" were not there, but we never would have known until we actually drilled and took a look around.

Since the above mentioned Gulf of Alaska well was drilled, other notable deep/expensive test wells have been drilled; some of them successful, others not. The so-called "Blackbeard" well was drilled off Louisiana's coast, in the Gulf of Mexico. Drilled in the Continental Shelf (the relatively shallow waters which extend around the U.S. mainland), Blackbeard bottomed out a little more than 30,000 feet deep, at a reported cost of over $200 million. It's now being extended even deeper, in an effort to find more deep deposits of natural gas for future supplies close to home. Blackbeard's owners have contracted the "Gorilla IV" (a great name, for a really big and powerful drilling rig) to move in and deepen the

Blackbeard well. Of note is the $190,000 *per day* it costs to hire Gorilla IV. That will work out to about $17 million for the expected 90 days of further drilling, plus still more for "completion costs."

Wind Power

Wind is a rapidly growing industry because it provides an energy source with no fuel cost and one which is environmentally attractive. Power providers can put turbines on a farm or a ranch and it doesn't bother the crops or the cattle, and the property owner earns extra income for rentals or royalties. The big issues are consistent power supply volumes (the wind does not always blow) and transmission facilities as noted above. Wind towers and turbines are expensive, but once in place relatively inexpensive to operate (no fuel cost). A significant problem with them is that each tower (each turbine) produces relatively little power, so it takes quite a few towers to produce enough power to make a significant contribution to the grid. Newer designs are becoming more productive, but their output is still measured by megawatts in single digits.

Development firms are now building numerous "wind farms" which will generate a great deal of power, but at present it's still a relatively-insignificant 1% of the overall U.S. power supply (about 20,000 MWh at this writing), though that represents an impressive five-fold increase in the past eighteen months). Mr. Pickens' plan is to increase that another 20x within ten years … *that's aggressive.*

Can we build the transmission grid to carry that power to markets, and can we get that many towers and turbines manufactured and up and running in that period of time? I doubt it. Which doesn't mean we shouldn't do it, just don't bet on its happening too quickly. We can't manufacture those turbines and towers and rebuild the transmission grid overnight. But building them will in fact create a lot of good jobs, and will make a big difference in the long run.

Hydrogen

Hydrogen-powered fuel cells are a much-discussed and very attractive alternative energy source. They need no source of fuel, hydrogen is everywhere. In fact, hydrogen is much harder to contain than it is to find. Fuel cell energy results from electrochemical processes which generates no heat and therefore has no exhaust emissions to worry about.

But there are big-time limitations with hydrogen fuel cells. Yes we can make them, and yes, they work. There are a number of universities and small enterprises pursuing these technologies for special and specific applications right now. But the basic problem — for now, anyway — is one of power *output* versus *weight*.

Fuel cells can't carry much of a load; that is, one fuel cell doesn't deliver much power. To get a useful power output, you have to put many cells together in a single package. But fuel cell packages, like lead-acid batteries used in most cars, are *heavy*. A fuel cell pack large enough to power a single 15 watt light bulb would weigh about 15 pounds. Imagine the weight required to power an automobile. This limitation spells extremely high cost to get significant fuel cell energy into the marketplace.

Ongoing research with materials, processes, et cetera will certainly improve on these limitations in the future. Some day fuel cells will be one of the solutions to our energy supply puzzle. But it will not be soon.

Another interesting technological advance has just been announced by one of the major U.S.-based oil companies. It reforms conventional fuels, like gasoline, into hydrogen through an on-board process. This could purportedly *increase* fuel economy by three or four times, while dramatically *reducing* emissions. This process is not currently in commercial application, but is another of the one-day-it-may-be technologies suggested by Thomas Friedman (*see* note 37, below).

Biofuels

Biofuels are those produced from the utilization of plants (corn, sugar cane, algae, and others) as feedstocks for processing into fuels. We all hear the stories of isolated (and illegal) stills in the backwoods where Pappy processed "corn squeezin's" into near-100% grain alcohol (or "white lightning"). Anyone who has even sniffed high-proof grain alcohol (much less tried it) will not be surprised to learn that it will readily burn. Hence grain alcohol is, in specifically formulated compounds (one of which is known as corn ethanol), a viable fuel for combustion engines.

At least one major domestic oil and gas firm has a significant pilot program under way in the application of algae to motor vehicle fuels. While commercial hard results are not yet readily available, some of the reputed algae-related test results offer stunning and outstanding potential. Reportedly one such process can "tailor" the strains of algae growth to fit whatever end fuel product is desired: diesel fuel, gasoline-substitutes, or other liquid fuels. And the yield per acre in terms of usable end product is said to be 1,000 times that of corn ethanol. Whatever solutions science can find we need and can use, in terms of what we know now, we will ideally get to the use of hydrogen fuel cells for vehicles and nuclear for much electric power generation sometime in the future, but nothing leads us to believe it will be the near future.

What we don't need is another stroke of genius such as that of using corn-based ethanol for automobile fuel. That Federal boondoggle (opinion) has cost U.S. consumers millions of dollars in terms of poorer gas mileage (fact) from their cars. It has cost all taxpayers in terms of subsidies to corn growers and/or processors of ethanol; it has harmed the environment through increased emissions required to process corn into ethanol; and it has impacted the whole world in terms of increased and inflated prices for corn-related food products.

It made a few Senators happy … you can probably guess whom (or at least from which states). At least a couple of agricultural firms have done very well from this effort, it also made a great many corn farmers happy with high prices for their crops. But then they had to buy their fertilizer and seed for the next crop, and found out how much *that* cost with demand increasing. An all-around mess, not to mention the obscene volumes of fresh water wasted in this process. Yet our government still subsidizes this absurdity.

Other forms of cellulosic fiber for ethanol may prove viable, but the consumption of water continues to be a very significant (and limiting) factor.

I have heard individuals in the ethanol business say that they now have hybrid corn plants that require only a small fraction of the water necessary to grow conventional corn, and that they have processes in place that use only about 3-4 gallons of water per gallon of corn ethanol they refine. I candidly doubt their claims, especially as the gentleman I heard speaking went on to say they use only as much water as is required to refine a gallon of gasoline (<1 gallon of water). My calculation is that is closer to 10 times as much, which still ignores the water consumed in growing the corn or other feedstock.

Solar Power

Solar power is arguably the best of all worlds, but it too has shortcomings. There are two basic applications of solar power; one is to directly generate heat, which can in turn be used to heat homes and buildings; the other is the direct generation of electrical power through the use of photovoltaic cells which is the rapidly rising technology where much research is currently focused in the solar world.

As an aside, in about 1977, at an awards dinner for the winner of the Houston Science & Engineering Fair (who happened to be my son), I heard one of the senior technology officers of Shell (Oil) Research Company comment that their development target on photovoltaic cells was to get the cost down to less than $1 per watt. Shell felt that at break-point commercially feasible applications of direct solar-to-electric conversion would be viable. I just read today that indeed that has happened — only 32 years later — and with a dollar that is probably worth about 10 cents of what it was then.

Present applications include the installation of heat-generating solar units on the roofs of commercial (or other) buildings. Thus utilizing that "free" energy to heat the building or generate its electrical power. Others are the direct generation of electrical power in commercial quantities or as "booster" power generated on various smaller sites which can be used on site or resold back into the local provider's grid.

According to proponents of solar power, a solar collection array of 100 miles square (that is 100 miles on each side, total area: 10,000 square miles) located in the southwestern U.S. desert could provide enough electricity for the entire country. That requires a very large piece of land. While much of our desert is relatively unpopulated and void of much vegetation or habitation, there would be material environmental impact on the flora and fauna which might exist on any such single-purpose dedicated tract of land. Another issue would be the transmission of that power to the densely populated regions of the country such as our east coast. Our present transmission grid cannot do it. To its credit, the current administration is attempting to improve that situation.

A seemingly out-of-character objection to such projects came recently from one of California's U.S. Senators complaining that such collection projects (albeit much smaller) in the Mojave Desert would be too unattractive. If you've ever seen that desert, you know it's not very attractive now. California doesn't seem to want offshore drilling, oil refining operations or much of anything to do with energy. Yet they want plenty of electrical power and low-emission automotive and truck fuels. To me, California has always seemed a little less logical than the rest of the country. Along that line, with all their fiscal problems in that State, they could surely benefit to the extent of billions in tax revenues by simply allowing offshore oil operations to resume at a reasonable level.

Tidal and Wave Power

Another source of great potential power, Tidal power is at present of no material scale on the world stage. There are, however, applications in place employing electricity generated from such energy.

Two types of tidal power classes are in use; those using the energy of moving water to power turbines, somewhat analogous to the use of wind to turn the blades in "wind farms." The other system uses the energy from the height differential between high and low tides, these are called "barrage" applications. The large issues here are environmental, and a paucity of viable sites around the world. The process has to capture the high tide behind some form of barrier (a dam or natural lagoon) then release it in a manner which allows use of the "falling water's" energy to drive turbines. A third (recent) development, "tidal stream generators" — which can draw energy from ocean currents, again similar to windmills — shows potential as yet another source of electrical power generation.

Such applications are logically limited to areas where tidal flows are at some speed. Prototypes of such systems are in place, but no commercial applications are in place, to my knowledge.

KNOWN RESERVES
ENERGY FOR RIGHT NOW (AND TOMORROW)

Coal

The U.S. has extensive coal reserves; probably 200 years' worth at current rates of production, and assuming no improvement in technology that might locate and recover still more.

Clean coal technologies can enable us to use that huge resource with a reduced impact on the environment.

Natural Gas

We have large reserves of natural gas and even greater undiscovered-yet-probable reserves. In the "years" equation, we have something like 105 years of "known" gas reserves at present use rates and much more in the "probable" category.

We'll have a new pipeline coming on-stream in a few years (projected completion: 2018) to deliver natural gas to the U.S. — via Canada — from the North Slope of Alaska. Gas has been "shut in" there for many years, as there was no practical way to transport it down to the lower 48 states. The plan is for 4 BCF of gas to move through that pipeline every year, for consumption in the U.S. and Canada.

Crude Oil

Future oil supply is a crapshoot, depending partly on chance and partly on actions both at home and abroad. Specifically, one must ask: (a) What can (and will) OPEC do; and (b) What will *we* do?

Matt Simmons, an acquaintance of mine from Houston, is experienced in the oil and gas financial community. In 2005, he wrote an interesting book entitled *Twilight in the Desert*.[34] In it, Simmons hypothesizes that even Saudi Arabia can't effectively increase its crude production much, if at all.

Saudi Arabia has positioned itself to the world as the "swing producer" for world oil supply. It has demonstrated it can increase production by two or three million barrels a day at any time; Conversely, it has shown a willingness to reduce production by similar amounts if necessary to achieve "equilibrium" in the world's supply/demand situation (and to keep the price about where they and OPEC want it). In other words, Saudi Arabia will increase output to meet strong demand, and is willing to take the hit financially for most of the necessary cutbacks when there's a glut in the market.

While there are those who disagree with Mr. Simmons, two respected senior executives with large international oil firms recently theorized that roughly 100 million barrels/day is the world's maximum potential oil production.

At the same time, I read about two very senior former Saudi Aramco officials who are at odds over that country's future production capacity. One says they are about "maxed out"; the other says they can produce 15 million barrels/day for decades. (Saudi Arabia has never yet achieved that 15 million barrel a day production level.)

---

[34] Simmons, M. *Twilight in the Desert: The Coming Saudi Oil Shock and the World Economy.* (Simmons 2005)

In his book, Simmons also goes on to make the point that we are going to effectively reach "peak oil" much sooner than most of those concerned with that matter predict; on a global basis, not just domestic.

Can they do it? Will they do it? It's difficult to know for certain. Saudi Arabia has massive financial commitments and obligations of their own, as do most OPEC countries, and depressed price are being felt by them all in terms of cash flow.

100 million barrels/day would be about 20% more than we're currently producing worldwide. We don't "run out" then, that's just the *peak* (before the world's major producing oil fields start to decline and the price *really* goes up.) Our domestic (U.S.) production has already declined significantly. Many of the U.S.' high potential oil and gas drilling areas (offshore and ANWR) are presently off-limits under Federal mandate. That ban is presently being reconsidered, it may be changed or allowed to expire, or not, but who knows what the outcome may be (or when)?

FUTURE DEMAND FOR OIL

Forecasting future oil demand has long been a necessary exercise for major producers (and consumers) of energy. Three recent forecasts project demand through 2030. None of those foresaw the current economic downturn which has affected the use of energy to a material extent, but even so their trends are probably still reasonably valid for the next two decades, assuming the worldwide recession will begin to correct itself in the next year or two.

Here's what two of those say:

(1) In its International Energy Outlook for 2008, the IEA assumes world oil production in year 2030 will be 104 million barrels per day (MBO/D).

(2) One major U.S oil company's outlook is close: "slightly under" 100 MBO/D worldwide requirement by 2030.

If Matt Simmons is correct, we can expect oil shortages and dramatically escalating prices for oil-derived energy products; gasoline, diesel, propane, etc. These changes are likely to occur well before the currently forecast two decades. His thesis of considerably less (than anticipated) supply holds for natural gas availability in the next two decades as well. In other words, it's quite possible the world will never see that 100 MBO/D level of production.

If, conversely, Dr. Yergin is right, we may see that 100 million-plus barrels/day level of production within the next decade. Indeed there have been some major discoveries of late in the waters off Brazil, off the Western coast of Africa, and in the Gulf of Mexico.

On the other side of the crude oil supply debate, another acquaintance of mine, Dr. Daniel Yergin, chairman of Cambridge Energy Research Associates (CERA) projected in 2008 that oil production would rise to 109 million barrels a day by 2014. He now suggests that 7.5 million barrels of that increase is at risk due to economic conditions affecting the price of oil.

For certain the U.S. has significant probable and recoverable oil and gas supplies in Alaska's Arctic National Wildlife Refuge (ANWR) ... but it's off-limits too. (Ask your congressman about that logic).

New technologies in the oil and gas industry also have substantially increased the ability to recover oil and gas from so-called "tight sands." This technology has also made feasible the recovery of significant volumes of oil from "known but previously unrecoverable" oil and especially gas deposits in the upper-middle plains and western states where oil and gas operations have long been underway.

Shale

Another huge potential domestic resource is an oil-like hydrocarbon found in underground shale beds, known colloquially as "shale oil."[35] But there is no current commercial U.S. production, and we are still feuding over whether to try to extract it all. That's a fight between the environmental movements, state and federal politicians, and the oil industry. There is not yet any proven effective technology to *economically* extract and process it commercially. I had the opportunity to see one of the early shale test projects in the 1960's, shortly after those research and test programs had started.

---

[35] The material in "shale oil" deposits is not really crude oil, it's "kerogen" (a mixture of organic chemical compounds). But it can be processed into a useful, oil-like energy source. Associated "shale gas" is yet another potential source of significance.

Shale bed production is being seriously accelerated today by some of the major oil firms and in some government and university laboratories. It is generally acknowledged that it will be proven economically practical, and that it can be done[36] ... *if* the environmental issues can be resolved. But there are significant environmental and water-usage impacts associated with shale development. That process is expensive, and today's low energy (oil) prices are not conducive to such investment.

Geothermal Energy

Geothermal energy is that which is extracted from the earth's stored heat. There are more than 20 countries which produce geothermal power, of those only five or six produce significant portions of their electricity from that source.

Chevron Corp is the largest producer of such energy in the world. It comes primarily from "the Geysers," a field in California (Chevron's home state).

Applications of geothermal energy are to drive turbines in "dry steam plants" or in" flash steam plants" which use high-pressure hot water pulled into low pressure tanks, resulting in flash steam used to drive the turbines. There are also binary cycle plants which can use lower temperature water employed with secondary fluid which can flash to vapor as a turbine drive process. Flash steam plants are most common in current operations with the binary cycle ones being more often built now.

Historically geothermal power plants have been limited to proximity with the edges of tectonic plates (where steam typically escapes).

Development of the "binary" plants gives hope to the practicality of drilling and employing geothermal energy over a larger geographical footprint. There are, however, issues with the eco-

---

[36] This, again, speaks to the world price of crude oil, which demand (and/or OPEC) can move up and down almost at will.

nomics of drilling the large-bore wells to the necessary depths for such applications. More broadly employed (by around 70 countries) is the direct application of geothermal energy for heating purposes.

The theoretical limits of geothermal to heat and power the earth are virtually limitless. The earth's heat content could last for several billions of years. Theory and practical application may have a long way to go. And such energy is not without its own environmental impact with the gases which are released from such activity.

In terms of scale, current geothermal applications are insignificant to worldwide or domestic energy consumption.

Water Resources

While we don't intuitively associate water with energy resources, fresh water is an essential component in much energy production. As mentioned above, ethanol requires inordinate amounts of water; indeed oil refineries utilize water (principally as a cooling element) in the production of gasoline and other fuels. Drilling wells for oil and gas requires water as a part of the drilling fluids mix.

While our Midwestern neighbors – who have suffered the ravages of flooding more than once in the past few years – probably believe there is far too much water in the U.S., there is in fact a shortage of water in much of the agricultural west. There it is becoming a more and more valuable commodity. We read about farmers who debate whether to plant crops which they need to irrigate, or to just sell their water rights to municipalities or other potential buyers. In California, Nevada, and Arizona, water becomes more precious year after year. More people, more crops, more need for irrigation, crops, lawns, and — in much of the desert — golf courses.

In California, the state was recently offering $275 per acre/foot to buy water. That's about 300,000 gallons of water. In that part of the country a suburban home will use around half that much water per year.

The point about introducing water into this discussion is that it's another commodity (and cost) which has to be considered in the process of resolving our energy puzzle.

THE PICKENS PLAN

Right now the same T. Boone Pickens mentioned above is aggressively pursuing a plan to convert a significant portion of our automotive and truck fuel to CNG (rather than gasoline) and shift much more electricity generation to Wind power.

To me, those are excellent ideas, but the best part about his advertising and pressing on with this Pickens Plan is the publicity it (and he) gets which may get us all exercised enough that we will require our politicians to seriously address the energy situation.

The two major challenges that I see facing the Pickens Plan are:

1. Unless and until we learn how to store massive amounts of electricity, we can't use wind turbine power to supply the "Base load" because when the wind doesn't blow to turn the blades, there is no power generation.

2. The shortage of transmission lines to move that wind-generated power from the central part of the U.S. (where the wind blows) to the coastal regions where larger populations consume that electricity.

A third issue is that cars and trucks have to be converted to run on CNG ... yet another significant individual investment.

There is quite possibly some creative genius in a laboratory or a board room right now figuring out how to store and/or transport electricity efficiently and in the volumes required. But that won't be developed overnight either. But when it is, somebody will get seriously wealthy. Similarly, someone in some other laboratory is breaking the mystery of the "super-battery" that can effectively generate commercially significant power on its own.

We can build the transmission lines, but not quickly and not without a great deal of litigation and delay over property rights, environmental matters, and the like ... along with a gigantic investment. We can retrofit our automobiles to burn CNG but that involves a significant ($2,000 or more) per car outlay, and you then need to find a CNG fuel facility to fuel it.

This infrastructure is a major issue. Delivering CNG to automobile drivers presents a huge challenge. We have fewer than 1,000 CNG fueling stations in the U.S. and about a fourth of those are in one state (California). There are thousands or millions of "normal" gasoline pumps in the U.S., but they can't be magically switched over to CNG delivery. That will take time and a great deal of money too. There are home fueling systems for CNG vehicles but few manufacturers make CNG cars (none currently in the U.S.). Research says there are some 8,000,000 CNG-fueled cars in the world, so it's obviously a viable transportation fuel.

Technology and invention are indeed wonderful things.

To that point, the Pulitzer Prize-winning author Thomas L. Friedman has recently published a book on his view of world conditions (*Hot, Flat, and Crowded: Why We Need a Green Revolution — and How It Can Renew America*).[37] In it, Friedman theorizes that solving U.S. energy problems could be accomplished via invention, innovation and technology.[38] I don't disagree, but when does that start to happen?

Speaking of Mr. Friedman: I recently heard a friendly and robust interchange between him and a well-known radio personality, a self-described "nonbeliever in the world of global warming."

---

[37] Originally published in 2008, an updated version of Friedman's book was released in paperback just as this book went to press, in November 2009.

[38] (Friedman 2008)

Friedman suggested then that they talk about the "Flat and Crowded" aspects of this recent book, omitting the "Hot" components.

In Friedman's vernacular, "Flat" speaks essentially to the fact that knowledge is now so widely available - thanks to the Internet and other instant communications technologies - that knowledge is shared worldwide on a real-time basis. In other words, there is no pinnacle of knowledge (which is assuredly true, in my view). A scientific breakthrough in a laboratory in the U.S. is known in China, India, and Timbuktu, virtually at the same time. All the world's knowledge is no longer confined to the more developed and progressive nations ... it is everywhere and, it gets there instantly.

Friedman said in this dialog that he felt the U.S. could, should, and would "innovate," not "regulate" its way out of the financial dilemma in which we find ourselves, especially as regards the energy and environmental issues. He went on, however, to propose that a "floor" price of $4/gallon for gasoline be imposed by the federal government in the form of a "gas tax." His rationale: that when gasoline prices reached nearly $4 at the pump in 2008, Americans dramatically cut back on their driving ... which indeed we did.

Friedman's thesis: If we realize gasoline will never be less than $4 we will reinforce the "drive less" habit. That's one possible outcome; the other is that we all just get accustomed to that price and "bake it into the cake" of monthly expenses, and drive, drive away.

That issue aside, concern of an overcrowded earth is reminiscent of the Club of Rome's "Limits to Growth" publication of the 1970s which incorporated dire predictions of famine and material shortages worldwide by the early 21st Century (right about now, actually). Yet it also predicted an economic collapse in the 21st century. Might we be in that one now?

As to crowding, no doubt the more densely populated countries are increasing at much greater rates than is the "western world."

Interestingly the Muslim nations and peoples located outside their native lands are experiencing a much higher birth rate than are the indigenous populations of their host countries. This may foretell a change in the balance of cultural and religious influence in many nations in the not-so-distant future.

THE PRICE OF ENERGY DEPENDENCY

I'm not a Conspiracy Theorist, but I do believe that people, nations, and organizations will act in what they perceive to be their own best interest. OPEC does, China does, India does, Venezuela does, Europe does, and so does the U.S. George Washington no doubt said it better: "No nation can be trusted farther than it is bound by its own interests." This is not likely to change, in my view.

The price we pay for our current oil addiction is huge in terms of dollars and it is significant to our economy and balance of payments. As stated at the outset: At $150 a barrel, that's three quarters of a trillion dollars per year. Even at today's much lower prices, our import bill is about a quarter trillion dollars. But that is not the greatest potential cost we face. Those issues are (a) security; and (b) political independence.

In the U.S. we tend to feel somewhat superior to much of the world, invulnerable, almost assuredly insulated from disaster ... but maybe we're not.

- What about 9/11?
- How about our current financial crisis?

Supply Risk

What if the Middle Eastern OPEC member states (along with Venezuela) literally cut off our supplies of imported oil? We'd still presumably have Canada and Mexico as reliable suppliers, but they would not be able to meet our daily needs (even at today's reduced levels of consumption). Between them, Mexico and Canada currently sell us about 3 million barrels daily. We import 10 million or more barrels from other countries, mostly OPEC members.

Without the OPEC countries' supplies, we'd have a serious shortfall, and our "Strategic Petroleum Reserve" has only about 60 days' supply in it, yet we can only extract that oil at a rate of some 4 and a half million barrels per day. So even if the government opened the taps wide, we'd be short 5 to 7 million barrels/day from Day One of a true oil embargo. Our military's capacity to respond to threats and to operate effectively would be badly impacted, and we'd all be in line at the gas stations.

The CIA has concerns over this supply interruption threat as well. George Tenet, former director of the CIA, notes in his book, *At the Center of the Storm*, that Bin Laden's goal for the 9/11 attack on the World Trade Center was the economic destruction of the United States.[39]

In the fall of 2008, for the first time, Russia — which is not, at the moment, an OPEC member nation — was represented at an OPEC Ministers' meeting. To our knowledge, OPEC and Russia are not exactly plotting against the U.S.; historically Russia and Saudi Arabia have not had strong diplomatic ties. But that could be changing. And what is our risk of a serious interdiction of supply from the world's two largest crude oil producers?

---

[39] (Tenet 2007)

Financial Risk

Another very real risk we face is that the U.S. could run out of *credit*. The Chinese government, with its immense financial reserve surplus, has become our number one lender. At our proposed new rate of government outlay we'll need a truly massive supplier of credit to purchase the bonds we will issue to cover our profligate ways. What happens when China says, "no more"? We can continue to print money, but if we do, it will rapidly become relatively worthless and inflation will run rampant.

Where We Stand Now

To summarize: We generate electricity from all presently-viable sources. The problem is with our consumption habits here at home. Emissions are a real issue; in the news today is the threat of global warming, which seems to be a fact (unless it isn't), and which points out the ways in which the world is heating up (unless it's actually cooling). Whichever; there are qualified scientists that can articulately argue either position, and the climate change's purported causes are neither simple nor definitive. For sure however, carbon-based fuels do create undesirable emissions, although their effect on climatic conditions is still under considerable debate.

For these reasons, it is imperative that the U.S. begin to develop its own capability to become much more energy self sufficient than we are now. We can do so ... and we *must*.

Marching Orders

What we must do — and I hope this treatise can contribute to that end — is recognize what *should* be done (and why), and what *can* be done practically, then insist that our government start to do those things. Let's forget about crazy objectives; we must set reasonable goals, things which *can be done in the near-term*, and recognize that the great rhetoric that we (the U.S.) can be "energy independent in ten years" is just conversation and wishful thinking. It's an attractive goal, but near a physical impossibility.

There are a few easy things you and I can start to talk about, write to our representatives about, and generally start to make some waves over.

As a practical matter, neither you nor I are likely to get on a plane tomorrow, fly to Washington, and walk into the foyer at 1700 K Street, NW, Suite 740; Washington, DC 20006. But there are much simpler ways to speak up. You can pick up the phone to call the local office(s) of your Representative and Senators. (You'll find phone numbers in your phone book, or find them online at www.senate.gov and www.house.gov, the official websites for both houses of Congress.)

There, in addition to phone numbers and addresses for local and Washington DC offices, you'll find email contact forms for every member of Congress ... including yours. With a few mouse clicks, you can write your Senators and your Representative, ask questions, and share your opinions.

Whatever method you prefer, it's important that you *make contact* somehow. Let your elected legislators know that YOU are concerned about energy, and that you want to know whether *they* are. Ask where he or she stands on one or all of these issues. If you talk to an actual person – probably a staffer or aide – he or she probably won't know (or won't say) what the Senator or Representative thinks about this issue or that issue. You can bet, however that all emails and phone calls from constituents are logged, sorted according to issue(s), and serve as a barometer of public sentiment back home. At the very least, the fact that you wrote or called, *and what's on your mind*, will get on "the call sheet." When more than a few stray calls or emails come in, expressing concern about a particular issue, someone on the office staff will start to look at that issue, and their boss will hear about it.

Does anyone *guarantee* that your Rep or Senators will leap into action based on your calls or letters? Of course not. But what *is* certain is that if you *don't* contact your legislators, they won't know what you think. In short, *if you and I don't do it, it won't happen.*

And soon – maybe not next month or next summer, but soon enough - we'll again be paying $4 or more for a gallon of gasoline. And guess whose fault *that* will be?

Some Easy Questions to Ask
Your Senator or Representative

These last two suggestions derive from the fact that Pickens has a lot of attention now, he's making waves, and we can help out! It may never be exactly what he proposes, but it's sure to reduce our present energy dependency on foreign crude oil.

- Do you support more nuclear power in the U.S.?
- Do you support drilling in ANWR?
- Do you support offshore leasing and drilling for oil and gas?
- Do you support clean-coal technology power plants?
- What is your position on oil shale development?
- Are you in favor of more Wind Power in the U.S.?
- Are you in favor of continuing dependence on imported energy?

WHERE DO WE GO FROM HERE?

Recently I was visiting with a colleague about this project and he asked my conclusions regarding future energy sources and supplies for the U.S. My response was, and is: "We'll have to use nuclear for electricity generation, and we must move to natural gas and/or biofuels for transportation."

We cannot get to these destinations overnight, and in the interim we need to pursue development of all potential alternatives. Fuel cells and biofuels will be part of the solution, but not soon in my estimation. Nor do I have any confidence in the viability of corn-based ethanol as a competitive vehicle fuel. Other ethanol products, perhaps.

We must also recognize we'll have to develop "clean coal" technology, and utilize both solar and wind-generated power. We'll need to continue the search for and development of domestic oil and gas supplies.

To give credit where it's due, the current Administration is at least talking about expanding the electric power transmission grid and developing effective alternative energy sources. The problems are that they propose to do this by subsidizing numerous alternative energies, which may get them developed, but at the end of development, and the end of the day... they will have to compete on the basis of cost comparison with other less costly fuels. Subsidies shouldn't last forever. (To be sure, the oil industry has also received its share of subsidies and tax benefits over the decades.)

The automotive industries, either those in the U.S. (if any remain) or those in the rest of the world, will have to develop more fuel and emission efficient vehicles ... at competitive prices.

MY FOUR GREAT FEARS

- After the recovery from today's world-wide recession, OPEC will effectively drop the price of crude oil, and America will collectively say, "Oh, I guess there's no problem after all." We'll then continue our addiction to, and dependence on, foreign oil. As I write this, oil demand and price are down dramatically (not due to OPEC this time, but rather to slowing demand in the weak U.S. and world economies). OPEC is currently reducing its production and reintroducing "quotas" among its members to keep prices propped up.

- We, the U.S. will over-zealously attack the Carbon Problem, at great cost and with a significant, if indirect, tax increase on all Americans. Passions may incite the government to pursue this course without the benefit of good solid science as to just how large and real this carbon emissions problem is.

- "Cap and Trade" carbon taxes will become a reality, effectively saying: "You can create carbon emissions, but only if you pay extra to do so." Then those (potentially enormous) amounts of tax revenues will go into Congress' "cookie jar" for politicians' pet projects of all descriptions. Current guesstimates are that tax revenues from this might amount to $60 to $70 billion dollars per year. We'll all pay that through increased prices for energy.

- The effective socialization and/or nationalization of the financial, automotive and energy industries, With no clear path to the re-privatization of these great and important national industries.

Forecasting Change

Recently I posed this question to a dozen friends and acquaintances, all of whom are senior and experienced executives in various areas of the energy industry (oil, gas, coal, power generation/distribution):

> *How much of our electrical power do you believe will come from "alternative" energy sources (other than nuclear) by 2030?*

The most optimistic reply I received was: "Twelve percent. Maybe thirteen."

I just don't think that's going to get us where we need to go. Do you?

AFTERWORD

Every writer who addresses current events faces the ongoing problem of *time*: The passage of time between writing and publication means that most anything you write will — at the time it is read — have changed.[40]

With that preface, I'll date this book with a specific reference: As of this writing, in 2009, the Congress, the President, the Treasury Department, the Federal Reserve have funded and all the pundits are still debating the effective bailout of Wall Street, and many other institutions as well. This was necessitated largely by the "credit crunch" which froze-up financial markets here and around the world. It remains to be seen how effectively – and to what extent – it will be implemented. But I have great faith in our country, its people, and its institutions (with the possible exception of Congress).

Of particular note is that this huge uproar originally concerned the outlay of $700 billion (sound familiar?). Yep, that's a lot of money, *but at $150 a barrel, it will be our oil-import bill every year.* Yet the Treasury, the Fed, and the Congress sit on their hands and backsides about that particular issue ... though they (especially

---

[40] The reporter for the daily newspaper faces this problem ... although certainly to a lesser extent than the author of a book, as the reporter has the chance to rewrite, correct, and update today's stories tomorrow.

Congress) do love to pontificate. (By the way: that 700 billion dollar bailout has now grown to well over $1 trillion. These numbers are, I realize, not really understandable to most of us. How much is a trillion dollars? *A BUNCH.*

It's disappointing to me when the talented and highly-paid TV talk show hosts on Sunday morning say "everyone knows offshore drilling won't bring us any more oil for 10 years." Who is "everyone"? In fact, 'they' *don't* know that. And, in fact, I know better. But however long it takes, every day we don't *start* more drilling offshore puts that help a day further away than it is now.

Then we have the esoteric vision of electric cars, which seems like a dandy idea … but where do we get the electricity to recharge them? Perhaps from a power plant … I'd guess so. And maybe that plant needs fuel: coal, oil, gas, or nuclear … yes, one of those, or perhaps wind or solar power. It's hard to know for certain … *but it has to come from somewhere.* Most electric cars don't save energy; they simply *transfer* the needed energy source to a different place. They can potentially help the environment, but not necessarily. It all depends on where the electricity comes from in the first place.

Hybrid cars have the potential to recharge their own batteries, but apparently not to the extent needed to run indefinitely without an occasional "plug-in" or a boost from the onboard gasoline engine. In fact, reports about Hybrid cars seem analogous to stories about sail boats vs. power boats (I don't own either kind): "Sail boats are great. They're quiet, and they use only the wind's natural energy." But sailboat owners report the need to run under engine power about 80% of the time. Granted, hybrid car engines are significantly smaller — they burn less fuel and produce proportionally fewer emissions — than those in gas-only cars. But the improvement is incremental, not radical.

My point here is not that electric cars and hybrids aren't *helpful*; it's that they represent only a small step on our much longer journey. Worthwhile? Probably. A cure-all for our transportation needs? Probably not.

ACKNOWLEDGEMENTS

I wish to acknowledge numerous information sources used in this compendium. My primary "go-to" for energy data is the Energy Information Administration (EIA), a branch of the U.S. Department of Energy. The EIA has extensive data on its website, http://www.eia.doe.gov. The database maintained by the International Energy Agency (IEA) at http://www.iea.org, is also a great resource. Other websites and numerous other resources are detailed in the "Works Cited" section below.

I'd also like to thank the many people who helped make this work possible. Several Professors at the Colorado School of Mines were willing to share time and knowledge, and several helpful voices at the National Renewable Energy Laboratory in Golden Colorado, added to my own knowledge and to the contents of this work. Many old friends and business associates (too numerous to mention, but you know who you are) were kind enough to share their experience and wisdom, and for that I'm sincerely grateful.

Additionally I thank my son Bryan, himself a published author, accomplished musician, geologist (OU-trained), patent attorney, and entrepreneur. In addition to designing the book's covers and internal layout, he contributed heavily to the "Environmental Concerns" chapter and spent countless hours editing, formatting, and critiquing my work. Without his efforts this book would not be in print.

Robert Harston

ABOUT THE AUTHOR

If you like to skip chapters in books, this is the one to bypass, as it largely concerns why I felt compelled and qualified to write this commentary.

I'm a 77-year-old native Oklahoman, a 1954 graduate of the University of Oklahoma. OU happens to have the oldest school of petroleum geology in the country. I took some geology courses there, but I'm not a geologist. I've been retired 17 years, and enjoy living in Colorado.

I've enjoyed a great deal of good fortune in my life, being born in this country, raised in a loving middle class family where culture and talent were respected and admired, values were all about "others," Boy Scouting was a good thing, attending church weekly was "what you did." Later on, fortunate opportunities and great mentors enabled my career.

My experience in the energy industry began in the 1960s when I joined a small information services firm in Denver, Colorado. That company — aptly named Petroleum Information Corporation (or "P.I.," as it was known in the trade) — focused on information about the petroleum industry. The firm was owned by four entrepreneurial brothers, two of whom came from oil industry backgrounds. P.I. compiled data petroleum industry data — such as which wells were drilling where and at what depths, geological formations encountered, etc. — which it published in multimedia reports. In the '60s we had begun computerizing those data bases for our larger customers.[41] Being a marketing type, I started as sales manager. Thanks to the talents, help and support of numerous colleagues, eighteen years later I was in the "corner office" as President and CEO. I retired in 1992.

Thanks to lucky timing, vision by the founders, and strategic acquisitions, we grew that business from revenues of about $1.5 million (U.S. only) to about $100 million (worldwide) by the early 1980s. In the interim I had the opportunity to work with many earth-science professionals in both large and small oil and gas firms, domestic and international as well, relocating between Denver and Houston a few times.

P.I. was acquired in the late 1960s by the Nielsen Company; Dun & Bradstreet (D&B) acquired Nielsen in the early 1980s. About that time, the U.S. oil industry suffered a severe downturn when OPEC lost control of the market due in part to disruption of production in Iran by the overthrow of the Shah. Many OPEC members "broke ranks" and overproduced their quotas, creating chaos in the oil markets. Crude prices fell by about two thirds down into the $20's. The U.S. industry reacted and exploration activity dropped like a stone. Our customers all cut back radically, our market shrunk, we downsized from 2,000 employees to half that number.

In 1990 D&B sold P.I. to a Houston-based venture capital group which subsequently sold it to a Denver based information services company. That firm now operates as IHS Energy and in 2005 was listed on the New York Stock Exchange.

To set the record straight and as to where I'm coming from: I love the free enterprise system, and American businesses, large and small. The energy industry was, is, and will continue to be, a major and vital component of our nation's fabric. I always found the men

---

[41] If you think this sounds like any other "typical" data-oriented, computer-based undertaking, you'd be right … except for one detail: This was *way* before the so-called Information Age began. In the 1960s, computerizing *anything* was a challenging and pioneering endeavor, requiring (literally) rooms full of computers to manage even the most basic database applications.

and women of that industry to be honest, professional, smart, dedicated, and considerate. They made it a fascinating industry in which to work.

The oil & gas industry's associations and professional societies with which we worked closely (and supported extensively) included the America Association of Petroleum Geologists (AAPG), the American Association of Petroleum Landmen (AAPL), The American Petroleum Institute (API), and the National Ocean Industries Association (NOIA).

Finding oil is neither easy nor cheap. Envision sticking a steel straw, six inches in diameter and a mile or two long, into the ground ... and finding what you hoped for. The search for oil and gas deposits makes that industry a voracious consumer of data, which made my career most interesting.

I enjoyed many opportunities to work closely with senior executives in the big "major" oil and gas firms, attend industry association functions and conventions, and work with earth scientists.

So why am I writing this? Because as Bob Dylan said, "The times, they are a-changin'," and so will our energy consumption and sources. I'm a reader, and the more I read, the more I realize that many of our citizens and much of the media simply do not understand energy supply or demand. Indeed the politicians do not understand it either, nor do most of them seem to want to. Perhaps this project will provide some factual wisdom on the subject, and make us a bit better able to judge our best future options for energy.

WORKS CITED

API Newsroom. *Oil, natural gas supports 9 million American jobs, 7.5 percent of GDP.* September 9, 2009. http://www.api.org/Newsroom/industry-supports.cfm (accessed December 8, 2009).

Boren, David. *A Letter to America.* Norman, Oklahoma: University of Oklahoma Press, 2008.

Central Intelligence Agency. *CIA - The World Factbook.* 2009. https://www.cia.gov/library/publications/the-world-factbook/index.html (accessed November 16, 2009).

Derrick Publishing. *Derrick's Hand Book of Petroleum.* Oil City, PA: Derrick Publishing Company, 1898.

European Commission, The. *EU at a Glance; European Countries.* 2009. http://europa.eu/abc/european_countries/index_en.htm (accessed November 16, 2009).

Friedman, Thomas L. *Hot, Flat, and Crowded: Why We Need a Green Revolution--and How It Can Renew America.* New York: Picador, 2008.

Mouawad, Jad. "Estimate Places Natural Gas Reserves 35% Higher." *The New York Times,* June 18, 2009: B1.

OPEC Secretariat. *World Oil Outlook 2009.* Annual Outlook, Vienna, Austria: Organization of the Petroleum Exporting Countries, 2009.

Plimer, Ian. *Heaven And Earth: Global Warming - The Missing Science*. Lanham, MD: Taylor Trade Publishing, 2009.

Potential Gas Committee. *Potential Supply of Natural Gas in the United States*. Biennial Assessment of Natural Gas Reserves, Boulder, CO: Colorado School of Mines, 2008.

Simmons, Matthew R. *Twilight in the Desert: The Coming Saudi Oil Shock and the World Economy*. Hoboken, NJ: John Wiley & Sons, Inc., 2005.

Tenet, George. *At the Center of the Storm: My Years at the CIA*. New York: HarperCollins Publishers, 2007.

Tertzakian, Peter. *A Thousand Barrels a Second: The Coming Oil Break Point and the Challenges Facing an Energy Dependent World*. New York: McGraw-Hill, 2006.

U.S. Energy Administration. *U.S. Natural Gas Prices*. November 30, 2009. http://tonto.eia.doe.gov/dnav/ng/ng_pri_sum_dcu_nus_m.htm (accessed December 8, 2009).

INDEX